Stephen Alfred Forbes

A Preliminary Report on the Aquatic Invertebrate Fauna of the Yellowstone National Park, Wyoming, and of the Flathead Region of Montana

Stephen Alfred Forbes

A Preliminary Report on the Aquatic Invertebrate Fauna of the Yellowstone National Park, Wyoming, and of the Flathead Region of Montana

ISBN/EAN: 9783337267087

Printed in Europe, USA, Canada, Australia, Japan

Cover: Foto ©berggeist007 / pixelio.de

More available books at **www.hansebooks.com**

[ARTICLE 6.—EXTRACTED FROM THE BULLETIN OF THE U. S. FISH COMMISSION FOR 1891. Pages 207 to 258. Plates XXXVII to XLII.]

A PRELIMINARY REPORT

ON THE

Aquatic Invertebrate Fauna of the Yellowstone National Park, Wyoming, and of the Flathead Region of Montana.

BY

S. A. FORBES,

PROFESSOR OF ZOOLOGY, UNIVERSITY OF ILLINOIS.

[Date of publication, April 29, 1893.]

WASHINGTON:
GOVERNMENT PRINTING OFFICE.
1893.

6.—A PRELIMINARY REPORT ON THE AQUATIC INVERTEBRATE FAUNA OF THE YELLOWSTONE NATIONAL PARK, WYOMING, AND OF THE FLATHEAD REGION OF MONTANA.

BY S. A. FORBES,

Professor of Zoölogy, University of Illinois.

INTRODUCTORY.

The immediate impulse to the investigation of the aquatic invertebrate fauna of Wyoming and Montana, here reported in a preliminary way, was supplied by the ichthyological work of Dr. David S. Jordan, in the Yellowstone National Park, in 1889, and of Prof. B. W. Evermann, in Montana and Wyoming, in 1891.

The waters of Yellowstone Park had been reconnoitered by Dr. Jordan for the special purpose of ascertaining precisely which of them were destitute of fish and what was the cause of their barrenness. This having proved to be topographical in every case—some physical barrier to the entrance of fishes from below—it seemed possible to stock these waters permanently with valuable game-fishes, and thus greatly to increase the attractiveness of the Park to a considerable class of travelers. Preliminary to this, however, it was evidently desirable that a full knowledge should be had of the variety and abundance of the lower animal life of these fishless waters, since upon this the fishes introduced must chiefly depend for food. To this practical end it was the wish of Hon. Marshall McDonald, United States Commissioner of Fish and Fisheries, that my own investigations made in 1890 should be immediately directed; but with the understanding that the opportunity thus afforded for a general zoölogical survey of the waters of Yellowstone Park should be improved to the best of my ability.

My associate in 1890 was Prof. Edwin Linton, of Washington and Jefferson College, Pennsylvania, who, although specially charged with another duty, that of a study of the parasites of fishes in these waters, rendered me constant and invaluable service in my own special field.

In 1891 it was my general purpose to coöperate with Prof. Evermann in an exploration of the waters of Montana and Wyoming, to be made with reference to the location of a fish-hatchery; but in this, as in the preceding year, I made every effort to become as thoroughly acquainted with the animal life of the waters which I examined as the brief time spent in each locality would permit.

207

TRIP OF 1890.

Leaving the University of Illinois July 11, I was joined in Chicago by Prof. Linton July 14, having spent the interval in supplying deficiencies in our outfit. We left Chicago on the evening of the 14th, reached Mammoth Hot Springs, in Yellowstone National Park, via the Northern Pacific Railroad from St. Paul, during the afternoon of July 17, and went into camp the same evening on Swan Lake Plateau, with everything ready for the field. Our party at starting consisted of Prof. Linton and myself, our guide, Mr. Elwood Hofer, who had made our camp ready in advance, and a teamster, two packers, and a cook. Our outfit contained (besides the necessary camp equipage, pack animals and saddle horses for six men, a portable canvas boat with two pairs of oars, two naturalist's dredges with ropes, a set of portable sieves for assorting the contents of the dredges, a sounding line, a very deep trammel net 50 yards long, a creek seine, an ordinary minnow seine, a Baird collecting seine, surface nets, hand nets, two deep-sea thermometers, a dissecting microscope, a compound microscope with complete equipment for field microscopy and for the preservation of perishable minute material, tanks of alcohol, bottles, vials, etc.

Breaking camp on the morning of July 18, we rode 25 miles through Norris Geyser Basin and down the Gibbon River to the branch of the latter known as Canyon Creek, where we camped for the night and made our first collections with hand nets from that stream. On the 19th we rode through the lower and upper geyser basins and camped just beyond the latter, on the banks of the Firehole River.

Collections with hand and surface nets were made here from various points on the Firehole and from the outlet of a warm spring on its banks. As we were now to travel for some weeks by mountain trails, the teamster was here turned back, and the pack animals were loaded for the trip across the "continental divide." Leaving this camp on July 20, we crossed the divide through Norris Pass and went into camp on the shore of the north end of Shoshone Lake, at the mouth of Heron Creek. A hurried dip with surface nets was made, in passing, into the waters of some large ponds, without outlet, in the mountains near the summit of the divide.

On Shoshone Lake we stayed for the three days following (two of the party circumnavigating it on the 22d), and made extensive collections along shore, in the inlet of the lake, in an overflow lagoon or pond beside it, and from its own waters with towing net and dredge, from the surface by day and night, and from the bottom at depths varying from 8 to 40 feet. Breaking camp on this lovely lake, which will ever have a peculiar charm in our memories as the place where systematic work on the invertebrate life of the waters of the Park began, we went on the morning of the 24th to Lewis Lake, 12 miles below, two of the party running the rapids of Lewis River in the boat. We camped on the east shore of Lewis Lake, working July 24 and 25 with dredge and small nets in the lake, and making miscellaneous collections from springs of various temperatures and from the waters of a swamp which becomes connected with the lake in spring.

From Lewis Lake we rode to Heart Lake, a distance of 7 or 8 miles along the foot of the Red Mountains. Arriving at noon of the 26th, we crossed Witch Creek and camped in a grove of pines above its mouth, not far from the base of Mount Sheridan, whose precipitous front was a maze of roaring streams supplied by the melting

snows on its upper slopes. The situation here is one of the most attractive in the
Park. Camping ground and feed are good, water is abundant and excellent, fish of
three kinds—trout, chub, and sucker—are plenty in the lake, and minnows can be
taken by the half bushel in the warm waters of Witch Creek. The place is absolutely
retired (there was not even a trail by the way we came) and quite off any line of travel
even proposed by the Park authorities. The lake is a gem of beauty, a fit companion
to the noble mountain, from whose heights a view of lakes and rivers and mountain
peaks and ranges may be had second to none in this part of the Rocky Mountains.

Our stay in this charming spot extended to nearly five days, all of which but one
were spent in continuous collecting from the lake and from the tributary already mentioned as Witch Creek. Violent winds made it difficult to work far from shore in our
light canvas boat, but, with the aid of a small raft made for the occasion, we got good
soundings and dredged successfully about a quarter of a mile out. Here, besides the
kinds of collecting already specified, we used our small seine in Witch Creek and the
trammel net in the lake, taking in the latter considerable quantities of all the larger
kinds of fish the lake supports, in places where the rocky bottom would have made
seining impossible even with a much more cumbrous apparatus.

We shifted camp on July 31 to the west bay of Yellowstone Lake, passing Rid-
dle Lake on our way and pitching our tents on the shore, a few rods above the Upper
Geyser Basin of this bay. Here a line of soundings was run out about 2,000 feet,
the dredge was hauled from the boat 1,000 feet from shore at a depth of 102 feet,
with a bottom temperature of 46°, and at various lesser depths near shore. Other
collections were made from the lake in the usual variety, and also from several of
the warm springs and their outlets. The first water birds were shot here for a
study (by Prof. Linton) of the relations of the fish-eating birds to the parasitism of
the trout, and descriptions were made of rotifers and protozoa which it was not
possible to preserve for later study. A short excursion from this camp gave us
access with the boat and our lighter apparatus to Duck Lake, a land-locked body of
water, formerly connected with Yellowstone Lake, but having now neither inlet nor
outlet at any season of the year. At the foot of Yellowstone Lake, where we arrived
August 5, our party was reorganized by the dismissal of the guide and pack train and
the engagement of a teamster and saddle horses for the remainder of the trip.

From this point we worked on the lower lake, on Yellowstone River at the outlet,
and on Pelican Creek and smaller tributaries, until the 12th of August, Prof. Linton
going for pelican to the head of the lake, in a skiff, on the 9th and 10th, with a.volun-
teer party from the lake hotel. Towing-net collections were made by this party not
far from the inlet. The dredge was run from a skiff off the landing, on this visit, at
a depth of 100 feet, and also in shallower water. Being unable to reach deeper
water for want of a line left at Norris Geyser Basin, and needing also other supplies
left there, for which I was unable to get transportation to the lake, we left the lake
for a time, starting to Norris Basin on the 12th. The 13th was spent at the Trout
Creek camp, collecting in waters of various temperatures from Alum Creek, above
and below the remarkable hot-spring basin through which this stream flows.

The occurrence of small trout in the upper course of this little creek seemed at
first a mystery, since they are found above the hot springs which boil up in its bed
for a distance of several rods, and so make its waters there altogether intolerable to
fish; but it finally appeared that when the streams are filled by melting snow in

spring and early summer the temperature is so reduced, even in its hottest part, as to make it possible for fish to pass.

On the 11th we unloaded our boat for a few hours' work on Mary Lake, a clear and pretty sheet of water lying near the summit of Mary Hill, at an elevation of 8,200 feet, and we thoroughly examined also the upper waters of Nez Percé Creek as we passed down the stream to the Lower Geyser Basin. The 15th we spent in collecting from the lower part of Nez Perce Creek and from the Magpie, its principal affluent, for a short distance above its mouth. Every condition was found suitable in these streams for the maintenance of fish, and a report to the Commissioner from the field to this effect was followed by a consignment of Von Behr trout, set free in the Nez Perce by Mr. Lucas, of the Commission.

August 16 was spent on the Firehole River, from the mouth of the Nez Perce to the junction of the Firehole and the Gibbon, and the 18th on the same stream above the Lower Geyser Basin, our collections ranging from the ford above Old Faithful to the Middle Geyser Basin below the Excelsior Geyser. Excellent opportunities were given here for a study of the effects of the geyser outflow on the animal life of the stream. Some hauls along shore with the surface net were made, in passing, from Goose Lake, near the lower basin.

On the 17th we collected at Cañon Creek again, and from the Gibbon River, at the mouth of this creek, and also above and below the falls of the Gibbon, well known as an impassable barrier to the movement of fishes up this stream. In our collections above and below falls in these rivers and creeks, it was my object to learn whether any other animal inhabitants of these streams were similarly excluded from their upper waters.

On the 20th we made a trip from Norris Geyser Basin to the Twin Lakes (in one of which whitefish had already been planted by a Fish Commission party), returning by a small bracelet without outlet, called the Lake of the Woods. We went thence to the Grand Cañon, collecting by the way from the Gibbon above Virginia Cascade, and from minor waters passed, and on the 22d made a trip to a lake nameless to the guides, but marked Grebe Lake by the geologists, and mapped as the source of the Gibbon. We carried boat, dredge, small seines, and our lighter collecting apparatus to this lake, and thoroughly overhauled it, as typical of its kind.

Returning to Yellowstone Lake on the 23d, we explored Pelican Creek on the 24th for several miles above its mouth, and on the 25th dredged from two rowboats at a depth of 105 feet, with a bottom temperature of 42.5°. Returning next day to the cañon, we collected from the lower course of Alum Creek, the upper part having been explored by us previously. On the 27th, sending the wagon to Mammoth Hot Springs by the traveled road, Prof. Linton and I took the trail down the river to Yancey's Ranch, crossing Mount Washburn, and making considerable collections from Tower Creek above the fall. On the 28th we went from Yancey's to the springs, stopping by the way at Lava Creek and Blacktail Deer Creek. On the 29th the usual collections were made from the Madison, at the crossing of the Cooke City road, and also from Swan Lake on the plateau of the same name, to which our boat and the usual collecting equipment were transported for us by Capt. Boutelle, U. S. A., acting superintendent of the Park.

On August 30 we closed the field work of this trip with collections from Glen Creek below the falls, from the Gardiner River at the mouth of Hot Creek, near Mammoth Hot Springs, and from a small lakelet among the hills, towards Gardiner.

The accumulations of the trip were made under 387 collection numbers, representing 43 localities.

Our work was limited substantially to the central park plateau, only that about Mammoth Hot Springs passing beyond the lava formations which cover the plateau everywhere to an unknown depth and noticeably affect, as we discovered, the animal life of its waters. The river systems investigated were those of the Gardiner, the Madison, and the Yellowstone, on the Atlantic side of the "continental divide," and of the Snake on the Pacific slope. The principal fishless waters examined were Shoshone and Lewis lakes, the Upper Gibbon and connected waters, the Firehole and its branches, Goose Lake, Twin Lakes, Swan Lake, and Tower Creek. The effects of geyser and hot-spring outflow were shown especially by collections made from the Firehole and from Alum Creek; and those of the occurrence of falls in the course of these mountain streams were shown especially by collections from the Gibbon and some of its tributaries. The highest elevation represented by our aquatic material was that of Mary Lake (8,200 feet) and that of a small lakelet near Norris Pass, not far from the same level. The greatest depth at which we dredged was 195 feet in Yellowstone Lake, although this depth was exceeded somewhat in the work of the following year. The altitude of this lake is 7,740 feet above the sea.

As material for a study of variations in biological condition, we obtained an abundance of specimens for a comparison of the system of life in lakes, ponds, rivers, and creeks where no fish are found with those in which only a single species occurs, and with those supporting from three to eight kinds of fishes.

The effect of the "continental divide" or watershed upon the distribution of aquatic animals is, of course, amply illustrated by our material; and this, taken in connection with materials gathered the following year from lower altitudes, should show something of the limitation of range of several species imposed by differences of elevation and the like. The influence of widely different geological conditions should likewise become manifest as we compare the animals of the waters of the Park plateau with those outside.

My warmest thanks are due to Capt. F. A. Boutelle, acting superintendent of the Park, who encouraged and aided our investigations in every possible way, and to our guide, Mr. Elwood Hofer, whose tireless energy and active personal interest in our operations were greatly in our favor. He was not only the guide and manager of our movements, but a most efficient volunteer assistant in camp and in the field.

TRIP OF 1891.

Leaving Champaign, Illinois, accompanied by my university assistant, Mr. H. S. Brode, on the afternoon of August 10, I arrived at Livingston, Montana, on the evening of the 13th, by way of Chicago and St. Paul, and proceeded thence to Helena, to which place our outfit had been shipped from Washington and Champaign. Delay in the arrival of part of the equipment made field work impracticable until the afternoon of the 17th, when we made our first collections from the Jocko River at Ravalli, on the Northern Pacific Railroad in western Montana.

My immediate object on this part of our trip was an investigation of Flathead Lake and its tributaries, and among these especially a small, very cold trout stream previously visited by Prof. Evermann, and noted as suitable for the supply of the proposed trout hatchery.

Flathead Lake itself offered a very interesting contrast to Yellowstone Lake, examined the year preceding, but with sufficient resemblance also to make comparison instructive. It was then commonly reached by stage from Ravalli through the Flathead reservation—a trip which we took, on August 18, going thence by steamer to Demersville on Flathead River, about 25 miles above the lake. We began our collections from this river on the evening of the same day, and worked here also on the 19th, collecting especially from bayous and backwaters. Through the kindness of Mr. H. W. Parchen, president of the Helena Board of Trade, and of his associates in a game and fishing club, I had the use of a small steam launch for the work on the lake—an indispensable advantage, without which we could have done only a little imperfect alongshore work. Accepting the cordial invitation of this club to make their club house our headquarters, we went thither from Demersville in the launch August 20, and made our first surface net collections in the afternoon of the same day. This club house is built upon a large bay at the upper end of the lake, partially sheltered from the rather violent winds prevailing, and yet containing water of sufficient depth to illustrate fairly the deep-water conditions of this lake. It represented also every variety of shore and bottom—sandy flats, weedy shallows, rocky shores, and gravelly banks—and had the further advantage, for our purpose, of giving ready access to a considerable tributary of the lake, named Swan River on the map, but locally known as the "Big Fork." Our collections here continued over the three following days, and included surface net work of all varieties, many alongshore collections, and several hauls with the dredge, made by aid of the launch, in water ranging from 80 to 162 feet in depth. Considerable collections were also made on Swan River, especially upon the rocky rapids a short distance above its mouth.

On the 24th we made a horseback trip to Swan Lake, 12 miles above our quarters, and spent several hours collecting with our smaller apparatus from the lower part of that lake and from a cold trout stream emptying into the river a short distance below.

On the 25th we went by the regular steamer to the foot of Flathead Lake, where we made such collections from this shallow southern end of the lake and from its outlet (the Cœur d'Alene) as a heavy storm would permit, finishing our work in this region on the 26th, and starting for Helena and Yellowstone Park. Our work in the Park was confined to the northeastern part —not visited in 1890—and to Yellowstone Lake, to which I went especially for a more thorough use of the dredge with the aid of the passenger steamer than I could make from skiffs the year preceding.

From Yancey's Ranch, which we reached on the 29th, we explored Slough Creek above the lower rapids, and some alkaline ponds near Baronette's Bridge, and on the 31st went up the East Fork of the Yellowstone to Soda Butte Station on the creek of the same name. Collections were made on the way from Amethyst Creek, and from the East Fork of the Yellowstone, where this creek empties into it. September 1 we spent near Soda Butte Station, at work in the creek and in Trout Lake 2 miles north of the "station." Returning to Yancey's September 2, we examined the overflow waters of the creek and searched the East Fork thoroughly at the Soda Butte bridge, and finished our collections from the river in that vicinity.

On the 3d and 4th we continued to the cañon and to the lake. The 5th was spent in making shore collections from Stevenson Island in Yellowstone Lake and in an examination of the small ponds and bayous of the island itself. On the forenoon of the 7th we finished our work on Yellowstone Lake by making three hauls of the dredge from the little steamer *Zillah* in the vicinity of Stevenson Island, at depths varying from 20 to 198 feet.

The work of the season closed, September 10, with collections made from two localities previously examined by Prof. Evermann with reference to establishing a fish-hatchery—Bridger Creek and a cold spring adjacent, near Bozeman, and some springs and small streams near Boteler's Ranch, just north of the Park. The return trip was made by the Northern Pacific Railroad, September 10 to 13.

The collections of this summer were made under 73 collection numbers, representing 23 localities.

Apart from the practical points aimed at, and the opportunity to further extend our knowledge of the aquatic life of Yellowstone Park, a region whose zoölogy must long have an exceptional interest, I value the results of this year's work chiefly as affording the means for a comparison of the animal life of two lakes so similar in many respects as Flathead and Yellowstone, and yet widely contrasted in altitude, in geological surroundings, and in topographical and geographical relations. It is, in my judgment, by a thorough examination and critical comparison of typical situations like these that the most interesting and immediately fruitful additions to zoölogical science are to be made in this field. I have only to wish that a longer stay on each of these lakes might have made possible a more minute and exhaustive study of their animal life and its relations to varying conditions of depth, bottom, temperature, season, weather, bionomic association, and the like.

DISCUSSION OF THE COLLECTIONS.

While the partial and, in most cases, merely preliminary way in which the material of these expeditions has as yet been studied makes any full discussion of the results impossible, it seems best that a report of progress should be made, presenting a summary review of the invertebrate life of these waters in the midsummer season, with descriptions or determinations of such new or particularly abundant and important kinds as have thus far been made out. Such a statement will include, in fact, the greater part of the economic results of immediate utility, and may be said, therefore, to fulfill the leading object of the work. This report may be most conveniently cast in geographical form, the life of each river system being separately discussed; but, for want of time to examine the entire mass of the collections, only a preliminary account of the fauna of the still waters visited, from temporary pools to Flathead and Yellowstone lakes, will be given at present.

The systems to which the various waters examined belong are those of Snake River and the Columbia on the Pacific slope, and of the Yellowstone and the upper Missouri on the Atlantic slope. The first is represented by collections made in the southwest part of Yellowstone Park, the second by those from the Flathead region, the third from the north and eastern parts of the Park and from the vicinity of Boteler's Ranch, and the fourth by those from the branches of the Madison in the central western part, and from Bridger Creek near Bozeman.

The collections now reported were made from the waters named in the following list: A mountain pond near Norris Pass, Shoshone Lake, Lewis Lake, Heart Lake, Yellowstone Lake and certain of its tributary waters, Duck Lake (near the Yellowstone), Mary Lake, Goose Lake, Twin Lakes, Lake of the Woods, Grebe Lake (at the head of Gibbon River), Swan Lake (Yellowstone Park), a lakelet near Mammoth Hot Springs, Trout Lake (near Soda Butte), small ponds in the Soda Butte bottoms, alkaline ponds near Baronette's bridge and several other scattered ponds, Flathead Lake, and Swan Lake (Montana).

THE SNAKE RIVER SYSTEM.

This system was reached only in its head waters, all a few miles from the low "continental divide," which sometimes separates only imperfectly the waters of the Pacific side of the continent from those of the Atlantic slope. Shoshone and Lewis lakes of this system are, respectively, about 1½ and 2 miles in a direct line beyond the crest of the divide, and Heart Lake is less than 4. From these lakes and from their tributaries all the collections made in this district were obtained, with the exception of a little group snatched with the hand net, while the pack train was passing, from a mountain pond near Norris Pass, on the Shoshone trail.

This pond was completely stagnant and filled with growing vegetation including filamentous and gelatinous algae and fallen timber. The collection contains little to indicate the elevation from which it came, but is of interest in comparison with the contents of the very different waters of Shoshone Lake, a few miles away and 400 feet below. In this pond I found a small spotted larval salamander, with well-developed hind legs already budded out, a considerable number of young insects, *Limnea*, *Chironomus*, and *Corethra*—larva and pupa, an amphipod crustacean *Allorchestes dentata*, and a great number of entomostraca. Among the latter were *Leptodora shoshone*,* two species of *Cyclops*, *Daphnia pulex*, an undetermined species of *Daphnia*, and a *Ceriodaphnia*. A black spring-tail (*Podurida*) and a wheel animalcule *Lacinularia socialis*, occurring abundantly in globular colonies, were the only other animals recognized in this preliminary examination.

Shoshone Lake.—Shoshone Lake is a lovely little body of clear blue water lying at the level of 7,740 feet—almost exactly that of Yellowstone Lake. It is shaped like a blotted T, with the stem, 7 miles long, extending north or east from the Geyser Basin, at the head of the lake, and the crosspiece, at the eastern end, about 4 miles in length. The stem reaches a width of 1½ miles, but narrows near the middle of the length of the lake to less than half a mile. This lake lies charmingly secluded in a bottom of the densely wooded mountains which surround it everywhere except to the southeast. It is at present accessible only by mountain trail from the Upper Geyser Basin, and has fortunately been omitted from the system of improved roads now being made for wagon

* Described on page 251.

travel. The shores are bold but not much broken, steepest on the south and west, where the 8,000-foot line runs a quarter to half a mile from the margin of the lake. On the northeast a peak about half a mile back rises to 8,600 feet, and others nearly as high lie not far north of the eastern end. To the south the Pitchstone Plateau lifts its black and forbidding mass—patched with snow all summer—to a height of nearly 9,000 feet; and to the southeast, 7 or 8 miles away, but seemingly less than half as far, the Red Mountains rise, culminating in Mount Sheridan, 10,200 feet above the sea. The rampart of hills surrounding the lake opens out on the northeast, where Heron Creek comes in; on the west, to form the valley of Shoshone Creek; on the south, where Moose Creek drains a swampy tract about 1 mile across and 5 miles long; and on the southeast, where the waters of the lake pass out through Lewis River. Some smaller tributaries empty at the Geyser Basin, on the eastern end; and a number of little rivulets, dry at times, drain the hills at various points. Near the mouths of the larger streams, ponds or small lagoons occur, connected with the lake at high water, and in midsummer thick with vegetation and swarming with animal life. The immediate shores are commonly rocky except for an occasional narrow beach of black volcanic gravel. There is little weedy water in this lake, the sandy bottom bearing at best a sparse growth of *Potamogeton* and plants of similar habit.

The only soundings made by us were in the north arm or bay of the lake, where depths of 40 and 50 feet were reached from a third to half way across the mouth of this bay, starting from the eastern side. The bottom at these depths varied from sand to soft mud, the latter without vegetation, the former with a growth of *Cladophora*.

Our camp was placed in a small grove on the flat at the mouth of Heron Creek, where we had at hand the creek itself and a small, very weedy, and very muddy lagoon, filled earlier with overflow waters, but then disconnected from the lake. Our collections were made chiefly in the north bay of the lake, but a few things were taken from the western end, and a few collected alongshore as we made our way to the outlet. In the north bay, besides making collections along shore and with hand nets in the shallow water, we hauled the surface net repeatedly from the boat, from 8 a. m. to 9 p. m., in both clear and rainy weather, and dredged at various depths from 6 to 40 feet. Our larger apparatus was useless, as there were no fish in this lake.[*]

---

[*] Here we first heard, while out on the lake in the bright still morning, the mysterious aërial sound for which this region is noted. It put me in mind of the vibrating clang of a harp lightly and rapidly touched high up above the tree tops, or the sound of many telegraph wires swinging regularly and rapidly in the wind, or, more rarely, of faintly-heard voices answering each other overhead. It begins softly in the remote distance, draws rapidly near with louder and louder throbs of sound, and dies away in the opposite distance; or it may seem to wander irregularly about, the whole passage lasting from a few seconds to half a minute or more. We heard it repeatedly and very distinctly here and at Yellowstone Lake, most frequently at the latter place. It is usually noticed on still, bright mornings not long after sunrise, and it is always louder at this time of day; but I heard it clearly, though faintly, once at noon when a stiff breeze was blowing. No scientific explanation of this really bewitching phenomenon has ever been published, although it has been several times referred to by travelers, who have ventured various crude guesses at its cause, varying from that commonest catch-all of the ignorant, "electricity," to the whistling of the wings of ducks and the noise of the "steamboat geyser." It seems to me to belong to the class of aërial echoes, but even on that supposition I can not account for the origin of the sound.

The assemblage of animals in this lake offered a peculiarly interesting subject of study, since it included practically no aquatic vertebrates. There were, of course, no fishes at all; we saw no turtles or water snakes, but a single frog, and only one small salamander. The dominant groups were insect larvæ, leeches, amphipod crustaceans, and entomostraca. By far the most abundant aquatic insects were caseworms (larval *Phryganeidæ*), mostly pupæ at the time of our visit. There were no crayfishes and no isopod or phyllopod crustaceans; but two amphipod genera (*Gammarus* and *Allorchestes*) were very abundant. The *Gammari*, represented by a single large and robust species (*Gammarus robustus* Sm.), were exceedingly common, creeping or swimming about, on or near the bottom, inshore, especially where collections of débris from the inlet rested in hollows of the rippled sand. They sometimes rose to the surface at night, where our towing net occasionally took a surprising number of them—at one haul almost nothing else. They occurred also abundantly in our deepest dredgings, in the lagoons examined and in the streams flowing into the lake.

This lake seemed, indeed, a paradise for the *Gammari*, containing an abundance of food for them, both animal and vegetable, fresh and in process of decay, and scarcely anything that fed upon them in turn. The commonest large leech (*Nephelis obscura*, var. *maculata*) feeds upon them to some extent, as I found by the dissection of two specimens; but another of these leeches voided a large horsefly larva (*Tabanus*). Their own food, if I may judge from that of seven specimens which I dissected, is quite varied, consisting of rotting vegetation (whose condition was shown by the mycelial threads running through it), of fresh algæ, and other green-plant substance, and of entomostraca (*Diaptomi* as far as seen). The stomachs of three contained, also, a noticeable amount of pollen grains of the pine. In three of the seven specimens examined large numbers of *Gregarinæ* infested the intestine. Their probable effect was shown by the fact that the intestine was empty in two of the parasitized specimens. Our *Gammarus* was thus practically at the head of the biological system of this lake, which was for it a royal domain where it was free to exact tribute of all, yielding scarcely anything itself in turn. The females at this time had their brood cavities loaded with young.

The entomostraca were principally a single species of copepod—a very large blood-red *Diaptomus*, new to science and here described as *D. shoshone*. This occurred in great numbers in several hauls with the surface net, and could usually be seen on a calm evening near the surface, where its tiny sportive leaps at the air kept the water microscopically agitated, as if by minute fish. Another *Diaptomus*, near *D. sicilis* and perhaps a variety of that species, occurred much more sparingly; and a third species of this genus, described on p. 252 as *D. lintoni*, was less frequently seen than either of the foregoing. There were a few species of *Cyclops* here, *C. serrulatus* Koch, *C. ajacis* Forbes, *C. minnilus* (new), and perhaps others; also, a *Cypris*, a *Bosmina*, a *Chydorus*, a *Daphnia* (*D. pulex*), and *Polyphemus pediculus*. So far as the crustacea were concerned, the lake was in practical possession of *Gammarus robustus* and *Diaptomus shoshone*.

The food of the latter little species was peculiar at the time of my visit, and the collections consequently give little idea of its usual function in the biological system of the lake. All of more than fifty specimens examined from several of the Shoshone Lake collections had fed freely, and often greedily, on the pollen grains of the pine. Only a single specimen dissected contained also some fragments of another entomostracan, among which were single antennal segments of a copepod, probably of its own species.

This fact is but an illustration of the dependence of the animals of a lake on the contributions made to its stores of food by the surrounding land. As a surprising number of fishes profit largely by the terrestial insects falling into the water, so this little copepod horde must live for some weeks, to a very large extent, on the pollen of the surrounding forest, relaxing the pressure, for a time, on the plant and animal life of the waters, which is doubtless their more usual food resource. Would the destruction of the forests here seriously diminish the number of *Diaptomi*, and thus lessen the food supply of the young of the trout with which this lake has lately been stocked?

The large leeches taken here occur throughout the Park in suitable situations, and have been noticed by earlier collectors at Yellowstone Lake. At the Shoshone station they were frequently seen in the clear shallow water, either swimming actively or creeping along the bottom. They are carnivorous leeches, as already mentioned, almost the only native enemies of the *Gammarus* worthy of notice in Shoshone Lake.

A frequent and interesting occurrence during our visit was the appearance at the sunny surface of the lake of a large dark-gray caddis fly (*Neuronia* sp.) freshly escaped from its pupal prison and flitting rapidly along with its imperfectly expanded wings, just on top of the water, going with speed directly for the shore. The number of these insects—caseworms in the larval state—was shown by the thousands of their empty cylindrical cases washed ashore. Larvæ, pupæ, and imagos were all common at the time of our stay. The case of this species is composed of thin, irregular pieces of vegetation (largely fragments of leaves and epidermis of water plants), or of chitinous plates of insects, eked out by filamentous algæ and other miscellaneous objects, all cemented and imbedded in the tough secretion from the salivary glands of the insect itself. On preparing to pupate, the larva closes the mouth of its tube by a coarse latticework of hardened mucus which protects the insect within, permitting at the same time the free access of water. Shoshone Lake, it need not be said, was an ideal place for the breeding of these caddis flies, since it contained no common carnivorous animal large enough to attack them.

Chironomus larvæ were common in these waters, and their pupæ, ready to emerge, appeared often in the surface net.

The mollusks taken were limited to a few specimens of a large, dark *Physa*, with an exceedingly thin and brittle shell, and a small, heavy *Pisidium*, with a few conspicuous lines of growth. These last were mostly empty shells, collected from the hollows of rippled sandy bottom, where they were readily seen as one floated over in a boat. An occasional dragon-fly larva (*Libellulidæ*), large larvæ of *Hydrophilus*, a very few hydrachnids, some slender annelids—undetermined as yet, and very difficult of preservation—and a considerable collection of digitate fragments of *Spongilla* are worthy of mention. So also is the scarcity of waterbugs, limited, indeed, in our collections from the open lake, to a single specimen each of *Notonecta* and *Corisa*.

The small brown lagoon or pond already mentioned as occurring near the lake was an example of a kind quite common along the lake borders of this region. It is separated from the lake itself only by a narrow strip of beach, and is largely filled with pond-lilies (*Nuphar*), grass, algæ, and the like, which grow out of a deep, soft ooze. There was little in the assemblage of animal forms of this place to suggest its elevation of more than 7,000 feet, unless the scarcity of mollusks and the higher *Crustacea* be so explained. A species of *Physa* and one of *Pisidium* were the only mollusks taken. Insects were represented chiefly by ephemerid larvæ, larvæ of *Culex*, *Chironomus*, and other small *Diptera*, caseworms, *Notonecta*, *Corisa*, *Agrion* larvæ, and larvæ of *Dytiscidæ*; *Hydrachnidæ* by a single species of scarlet water-spider; *Amphipoda* by the lake species *Gammarus robustus* and *Allorchestes dentata*; entomostraca most abundantly by a *Daphnia* of pale pink color, not seen by us before, and here described as *D. angulifera*, by *D. pulex* in moderate numbers, by *Polyphemus*, *Scapholeberis mucronatus*, *Eurycercus lamellatus*, *Chydorus*, *Cypris*, *Cyclops gyrinus*, *C. serrulatus*, etc., and by no *Diaptomi*, so far as observed. Leeches were present, although not numerous—the species already mentioned (*Nephelis maculata*) and one not detected in the lake, *Aulostoma lacustris* Leidy. This pond thus differed from the lake in the larger number and variety of insects, especially in the larval state, by the absence of *Diaptomi*, and by the vast predominance of the new *Daphnia*. The latter had evidently been very much more abundant earlier in the season, as shown by the quantities of its summer eggs. These formed a film over many square feet of the surface and had been washed ashore in quantity as a scum-like deposit along the bank. A few of the females were still bearing their ephippia.

The collections made by Prof. Linton from the lagoon at the western end of the lake are similar, as far as they go, but contain no entomostraca.

Lewis Lake.—Lewis Lake is so closely associated with Shoshone that the two might very well be treated by the biologist as one. The water, shores, bottom, and surrounding country have substantially the same characters for both, and their free connection by a river without falls and only some 3 miles in length tends to obliterate any small local differences. The fact that fishes are excluded from both lakes by falls in their common outlet still further assimilates them in biological condition, the only noticeable differences remaining being those of size and depth.

Lewis Lake is but 3 miles long by 2½ in breadth, with a greatest depth, in our soundings, of 80 feet. It is rudely triangular in form—more distinctly heart shaped, in fact, than Heart Lake itself. Its level, 7,720 feet, is but 20 feet below that of Shoshone Lake. Its western banks are highest and boldest, the 8,000 foot contour running usually from a quarter to half a mile from the shore. On the north and northeast the country is relatively low toward Shoshone and Yellowstone lakes, and the immediate banks are occasionally bluffy and the shores are everywhere wooded. The Red Mountains are close at hand, a range of ten peaks to the southeast; and to the south looms the great Teton group, the noblest mountain view to be had from any part of the Park.

There is a small hot-spring basin at the northwest angle of Lewis Lake, and a swampy tract about half a mile square lies beside it to the northwest, connected with it for a fortnight or so during high water. At other times communication is prevented by a narrow strip of beach, sand and gravel, a few feet across. At a little trouble and expense a permanent passage-way for fishes might be made and maintained, giving free access to considerable breeding grounds and stores of food.

On the west side a small permanent creek came in, about 30 feet wide and 2 feet deep at the time of our visit, with several acres of somewhat swampy ground at its mouth; and here also were three small warm creeks (150° F.) and one of cold water (50°), the latter apparently supplied by melting snow from the borders of the Pitchstone Plateau. A series of small lagoons, filled with sedge and bulrush, open into the lake along this shore. The cold waters from the Red Mountain range are diverted from this lake by the course of Aster Creek, which drains the northern slope and empties into Lewis River, about 2 miles below the lake.

Treacherous and stormy weather during our brief stay prevented our making many collections in the open water. The dredge was hauled, in fact, but once, and then at a depth of 56 feet, about a quarter of a mile above the outlet; and even the surface net was only twice used far enough from shore to give us the so-called pelagic entomostraca. The remaining collections were gathered inshore, in the swamp adjoining, and from streams, both warm and cool, on the western side.

In the dredge, at 56 feet, with a bottom of fine black mud and dead vegetable débris, we took quantities of the large *Gammarus*, a few *Allorchestes*, many *Chironomus* larvæ, specimens of *Pisidium*, and an undetermined annelid not preserved in condition for identification. The preponderance of *Gammarus* was as noticeable here as in the companion lake, although fewer were seen along shore in shallow water. The same may be said concerning the *Diaptomi* taken in open water in the surface net. The gigantic *Diaptomus shoshone* was relatively much less numerous, however, than the much smaller *D. sicilis* var. It was feeding extensively upon pine pollen here, as in the other lake. Although properly pelagic forms—those most at home, that is, in the open water and in the deeper parts of the lake and found rarely, if at all, in the small lagoons—these *Diaptomi* nevertheless extend their range close inshore, where they might be seen with the naked eye in the water or taken in the net, even when the surface was decidedly rough. These collections contained many examples of a peculiar entomostracan (*Holopedium gibberum*) not noticed in Shoshone Lake, and also an abundance of a colonial rotifer belonging to the genus *Conochilus*—allied to *C. volvox*, but apparently undescribed. [*]

In the lake, near the entrance of the warm stream on the western side, were quantities of young water-bugs (*Notonecta*), an occasional *Corisa, Gammarus, Chironomus*, and ephemerid larvæ. Caseworms were also abundant in the lake, and the air alongshore was full of two species, one black and the other pale brown, just emerged and pairing.

From a small lagoon filled with the overflow waters of a geyser, at the western side, we took with the hand net several specimens of *Gammarus* and hydrophilid larvæ, a few *Allorchestes dentata* and young *Corisa*, many libellulid larvæ, large and small, and larvæ of *Agrioninæ*, a single small *Physa*, and several water-beetles (*Colambus*).

From a warm stream at the same place (150° F.) we have many ephemerid larvæ, several caseworms and a single hydrachnid, a great quantity of large and vigorous specimens of *Gammarus robustus*, and a smaller number of *Allorchestes dentata*, many examples of *Pisidium*, a *Physa*, a few annelids (*Oligochæta*), and a single leech (*Nephelis 4-striata*) not taken by us before. The *Gammarus* was feeding very freely on dead and decaying vegetation and filamentous algæ, with some fresh vegetable fragments and a little pine pollen. No traces of animal food occurred in two specimens dissected, one half-grown and the other of the largest size.

---

[*] Described on page 256 as *C. leptopus*, n. s.

From the swampy tract on the eastern side of the lake already mentioned collections were made by hauling a surface net in the open water, by searching dead leaves, and by washing off the lily pads (*Nuphar*) in the net. These waters were swarming with life, chiefly insect larvae and crustaceans. They apparently contained relatively few mollusks, several specimens of a large *Physa*, a few *Pisidium* and one *Amnicola* occurring in our collections. The insect larvae included *Agrion*, *Ephemera*, *Chironomus*, *Corisa*, *Hydrophilus*, and *Corethra*. The only amphipod crustacean was *Allorchestes dentata*, represented by but few examples, but the open water contained a great quantity of *Daphnia angulifera* and an occasional *Sida crystallina*. Among the lily pads the same *Daphnia* occurred, together with a great number of *Sida*, a few examples of *Cyclops* and *Diaptomus*, and several of *Daphnella* and of *Polyphemus pediculus* (young and adult), and a single short, dark cypris. Leeches and their capsules were frequent, the usual spotted and striate species (*Nephelis maculata* and *N. lateralis*). A single specimen of *Clepsine elegans* and another *Clepsine* not determined were also noticed. The capsules of these leeches were common on the leaves of water plants. Among less conspicuous objects, small *Hydrachnidæ*, *Hydra fusca*, and the colonial rotifer *Conochilus* were abundant.

If we may pause now to glance at the animal life of these three lakes, characteristic as they are for their region, as compared with that of similar lakes of much lower altitude—Lake Geneva, in Wisconsin, for example—we find that the large and conspicuous differences, so far as invertebrates are concerned, lie mostly in mollusks and crustaceans. The complete absence of *Unionidæ*, of *Paludinidæ*, *Melaniidæ*, and of *Valvata*, and the scarcity of *Planorbis* and *Amnicola* are cases in point. The absence of crayfishes, of *Epischura*, and of *Simocephalus* is the most notable distinction in the crustacean list. *Polyzoa* also were extremely few.

Heart Lake.—Heart Lake had to us the very especial interest that it gave an opportunity, hitherto unparalleled in this country, to study by comparison the effect of the presence of fishes on the bionomic system of a mountain lake; and as the barren waters of Lewis and Shoshone lakes have since been stocked with trout by the U. S. Fish Commission, the results of this comparison of native conditions may hereafter be checked and supplemented by a study of the later state of invertebrate life in these two lakes. This lake is situated similarly to Lewis and Shoshone, is of nearly the same size as Lewis Lake, and is in most respects a companion to that and Shoshone, but differs totally in the fact that its outlet is much frequented by trout and that it is consequently well supplied with fishes. It lies only 5½ miles from Lewis Lake, in a straight line, and about 6 miles from the southern arm of Yellowstone Lake, but the latter is on the opposite side of the divide and is consequently connected with a different system of waters. It is divided by a peninsula into two unequal parts, the larger of which, rudely rhomboidal in shape, is approximately 1½ by 2 miles and another. The smaller part is subtriangular, with principal diameters of about a mile and the narrow neck uniting these two divisions of the lake is about a quarter of a mile across.

Heart Lake differs from Lewis and Shoshone by its close proximity to the Red Mountains, especially to Mount Sheridan, and consequently by the much greater amount of snow water which it receives. At the time of our visit, during the last days of July, the rush of rivulets down the mountain slope, supplied by the melting snows, filled the air all day with a noise like that of a train of cars. This lake has also its hot spring and geyser basin, but receives through Witch Creek a relatively larger body of warm water than either of the others. It drains, according to the published map of the Geological Survey, a larger basin in proportion to its size and is bordered on the north by a marshy tract 2½ miles long by nearly a mile wide. Its surface lies 250 feet below that of Lewis Lake and 270 below the Yellowstone. Its waters are very clear, but are nevertheless much more weedy alongshore than those of either of the other lakes.

The slope of Mount Sheridan continues downwards into the lake a little distance, and the water consequently deepens rapidly from the eastern shore. About 200 feet out the depth was 94 feet; at 400 feet it was 124; and at 1,000 feet it was 146. The bottom temperature at this latter depth was 40° F.

Our camp was pitched on the western side, about half a mile from the mouth of Witch Creek, and our work was confined to this shore and to a distance of about half a mile along the northern shore. Our dredgings here were made in three localities: in shallow water inshore, at a depth of about 10 feet; upon rocks a little distance out, at a depth of 30 feet; and in deep water from 46 to 120 feet, with a bottom of soft mud. Collections were made with the surface net from the open water at various hours of the day from 9 a. m. to 9 p. m., under such conditions of weather as offered themselves, and also from shallow water among weeds, commonly near the bottom. In addition to these, considerable collections of fishes were made with the trammel net and the smaller seines, the latter of which we used in Witch Creek as well as alongshore in the lake itself, and from these fishes a quantity of material was obtained for a study of the food of the various kinds.

As might be supposed, some noticeable differences appear on a comparison of our collection lists, some readily accounted for and others at present inexplicable unless as the secondary or more remote effects of the first. It is naturally to be expected that in so small a lake, and one with so few opportunities for successful concealment or escape, the kinds of invertebrates on which fishes feed by preference would be unable to maintain themselves in as large numbers as in similar situations where fishes do not occur at all; and especially will this necessarily be true if we find that the fishes destroying these invertebrates are not strictly dependent on them for food, but eat other things as well. This is true of both the trout and the sucker, the former being almost indiscriminately carnivorous, and the latter mixing insect larvæ and the like with a large proportion of vegetable food.

It is probably in this way that we are to explain the fact that we did not find in our stay on this lake a single larva of *Neuronia* (the largest caseworm in these waters), so abundant in Shoshone Lake, nor a single amphipod crustacean (*Gammarus* or *Allorchestes*)—all large enough to afford an attractive food to one or all of the fishes in these waters. That they occur here I can scarcely doubt, although the distribution of the *Gammarus* seems at best very whimsical in this region, but they certainly were far less common than in the adjacent lakes. The absence of the larger leeches (*Nephelis maculatus*) may be due to our failure to find suitable places for them, or they also may be eaten by fishes.

More difficult to understand is the very remarkable fact that we did not find here so much as a single specimen of the almost gigantic copepod, *Diaptomus shoshone*, although its companion elsewhere, the smaller species of *Diaptomus*, was extremely abundant in all our open-water hauls. Equally difficult of explanation was the vast abundance of the entomostracan *Daphnella brachyura*—not once taken before we reached this lake, and here the most abundant species next to *Diaptomus*. With it was *Leptodora hyalina*, also extraordinarily common (for a predaceous species) in the product of every haul. The absence of *Polyphemus pediculus* is less remarkable, since it was not common in the other lakes, and the fact that *Holopedium gibberum* should be wanting here while common in Lewis Lake is without special meaning, since we failed to find it in Shoshone Lake.

A fuller discussion of this matter must be postponed until our materials illustrating the food of fishes taken here have been completely studied; and I will merely add at present some further details concerning the general collections.

The beach on the west shore was gravelly, in some places with boulders of considerable size, and occasionally with a stretch of sand. The bottom was covered with a growth of weeds (*Potamogeton*, algæ, etc.) and was greatly cumbered with driftwood. Our deepest dredging was made off this shore, beginning at a depth of 120 feet, with a temperature of 40° F., and ending at 46 feet, with a temperature of 53°. This haul, made from a raft and boat together, was about 100 yards in length, over soft mud containing some very fine sand, but consisting largely of organic debris, both vegetable and animal. The latter, minutely examined, was seen to be made up of the valves of entomostraca, fragments of the cuticle of insect larvæ, and the shells of rhizopods (*Difflugia* and *Echinopyxis*), while the vegetable remains consisted of minute pieces of *Vaucheria* and other filamentous algæ, and fragments of higher plants, with a multitude of shells of diatoms. The living animals of this haul were *Chironomus* larvæ and pupæ only, the former red, and many of them in their usual tubes composed of mud and minute organic remains.

The shallower hauls, made at a depth of 30 feet on a bottom of rock and gravel covered largely with *Cladophora*, aggregated about 50 yards in length. A quantity of material was brought up and all critically examined. It was composed almost wholly of red *Chironomus* larvæ and their tubes, together with a few specimens of *Pisidium*. A single ephemerid larva (*Cœnis*) was the only other animal found. On the various alongshore hauls, at or near the bottom and through the weeds, the following forms were obtained: Larvæ of *Chironomus*, ephemerid larvæ, caseworms (a single specimen seen), small hydrachnids, *Physa* and *Pisidium* (only a few of each), *Cyclops gyrinus* and other species, *Diaptomus sicilis* var., *Daphnella brachyura*, *Daphnia arcuata*, *Eurycercus lamellatus*, *Acroperus leucocephalus*, *Leptodora hyalina*, *Bosmina longirostris* and some undetermined copepods, *Stylaria lacustris* and other annelids, and an abundance of *Hydra fusca* of the brick-red variety. The most abundant thing was *Daphnella brachyura*, and the next *Diaptomus*. The *Cyclops* was common, as were likewise ephemerid larvæ and the small bivalve entomostracans *Eurycercus* and *Acroperus*.

The towing-net collections, made in the open water some little distance from the shore, contained a much smaller variety, all entomostraca and hydrachnids. The latter were few in number, noticed only in a single haul. By far the most abundant species were *Diaptomus sicilis* (?) and *Daphnella brachyura*, sometimes one predominating and sometimes the other. With these were *Leptodora* and an occasional *Daphnia*.

The abundance of fishes in the lake was shown by the fact that our trammel net, simply stretched in the open water in the evening and lifted at noon the following day, contained 87 fishes, 12 of them trout, the remainder suckers and chubs, the latter most numerous. At another setting of this net, near the mouth of Witch Creek, in 8 to 10 feet of water, 10 trout, 2 chubs, and 65 suckers were taken in six hours.

YELLOWSTONE RIVER SYSTEM.

Yellowstone River drains all the eastern and northern side of the National Park, more than half its area, and from these waters much the larger part of our collections was taken. Yellowstone Lake was visited both years; Pelican Creek and smaller tributaries at the northern end were searched; and Yellowstone River was examined at intervals from the lake to the mouth of the Lamar or "east fork." The smaller tributaries of this system examined were Alum Creek, Tower Creek, Slough Creek, Lamar River, Amethyst Creek, Soda Butte Creek, Blacktail Deer Creek, Lava Creek, Glen Creek, and Gardiner River. Collections were also made from numerous lakes and ponds connected with this drainage system: Duck Lake near the west bay of Yellowstone Lake; some alkaline ponds near Baronette's Bridge across the Yellowstone; Fish Lake, near the Soda Butte; Twin Lakes, on the flat dividing the head waters of the Gibbon from those of the Gardiner and drained by Obsidian Creek; Lake of the Woods; Swan Lake, draining into Glen Creek; a small lakelet near Mammoth Hot Springs, connected with the Gardiner; and Boteler Springs, outside the Park.

Yellowstone Lake.—With Yellowstone Lake we reach the aquatic headquarters of this region, the real center of interest and importance for the study of the invertebrate life of Yellowstone Park. It is the largest lake so near the summit of the Rocky Mountains, and, excepting its high altitude, presents every feature suitable to the maintenance of an abundance of animal life. Its zoölogical interest is fittingly supported by its geographic and scenic features, which supply an admirable setting to the picture of life that slowly shapes itself in the mind of the zoölogist as he studies its waters and their contents and the inhabitants of its bottom and shores in their relation to each other and to surrounding nature.

The geology of the region shows that the present lake is only the relatively small remnant of a much larger body of water which formerly filled Hayden Valley and extended down the Yellowstone nearly to the present falls; but there is, I think, no reason to believe that it has dwindled in zoölogical importance as it has in size. Except for changes of climate, the variety of animal forms a lake of this size may maintain need not be surpassed (and commonly is not) by that to be found in one many times its size. It is not likely that there was ever here, when this lake was largest and deepest, a special interior and deep water fauna, such as occurs, for example, in the Great Lakes of North America; for, if there were, remnants of it would certainly continue and would have appeared in our deep-water dredgings. As a home of animal life it has probably been for ages similar to what it is now, except that we must suppose that the single species of fish which now inhabits it—evidently an immigrant across the continental divide—has produced certain changes in the balance of life, some of which will doubtless become more apparent as our collections from this lake are thoroughly studied.

The most striking feature of Yellowstone Lake is the irregularity of its form and the consequent length of its shore line, such that with an area of about 140 square

miles, its shore line approaches 100 miles in length; a fact whose biological signifi-
cance will be understood if we call to mind that the greater part of the food of
fishes in such a lake is to be found among the weeds of its shoal waters alongshore.
From a main body of fairly regular parallelogram, 7 by 12 miles several large arms or
bays project to the west and south, giving the whole lake an extreme length of 20
miles and a width of 14.   The immediate banks are generally abrupt, although not
high, with a narrow gravel beach, and except along the tributary streams there is
little or no swampy ground about the shore. The Yellowstone above the lake, however, runs, as do most streams in this region, through a wide, swampy bottom, which
must afford an immense field for the breeding of fishes and for the growth of their food;
and Pelican Creek, the next largest tributary, is similarly situated, while several of the
smaller creeks have each at their mouth a little bar, piled up by waves and ice, which
has partly stopped the outlet and so formed above it a weedy lagoon or marshy bay.

The bottom commonly slopes gradually downward, making an abrupt descent, so
far as known, only from the shores of Frank Island and from the lofty summits about
Southeast Bay. Beyond the beach of gravel and boulders which commonly borders
the lake, comes usually a belt of sand or sandy gravel, and beyond this a sandy mud,
becoming finer and darker inwards, until in the deepest water reached it was a very fine
black ooze. The greatest depth reported by the Hayden survey was 300 feet, in the
center of West Bay. My own deepest soundings were incidental to our dredging operations, and were limited to a distance of 2,000 feet from shore off our camp at Hot
Spring Basin on West Bay, and to an area north and east of Stevenson Island, where
the lead was dropped at distances varying from half a mile to 2 miles from the island.
The greatest depth reached in this area was 231 feet, at a point nearly equidistant
from Stevenson Island and Steamboat Point.

My only temperature observations were made August 4, 1890, at which time the
surface temperature of the water was 62° F., the bottom at 100 feet was 46°, and at
184 feet, 42½°.

This lake lies almost at the summit of the Rocky Mountain watershed, the continental divide following approximately the outline of its western and southern borders
for about 40 miles, at distances varying from a mile to 5 miles from the shore. To the
westward of the lake lie broken pine-covered hills, which rise from 250 to 800 feet
above its level. To the north are the dark ridge of the Elephant's Back, about 850 feet
above the lake, and the Sulphur Hills, which finally rise to a height of 9,000 feet above
the sea. On the east lies a mass of rugged volcanic mountains, a part of the Absaroka
Range, patched with snow all summer. They approach the shore most closely along
the southeast arm at the upper end, where the scenery is very bold and fine. A few
peaks rise to a height of 11,000 feet above the sea at a distance of 5 or 6 miles from
the shore. The boldest elevations are those just below the mouth of the inlet, where
mountains less than 2 miles away reach a height of 9,600 feet.

Besides the Upper Yellowstone and Pelican Creek, already mentioned, the principal
tributaries to the lake are a number of small streams which drain these eastern mountains, and, taking their waters from the melting snows and flowing most of their way
through overhanging forests, bring to the lake a considerable amount of very cold
water. The hot springs and geysers are found mostly on the western arm and at
Steamboat Point in the northeastern part of the lake, but the amount of warm water
contributed by them is quite insignificant for so large a lake.

In the open water there was always a very fair supply of entomostraca, both Cladocera and Diaptomi, but at the time of our arrival on the West Bay the phenomenal fact was the vast abundance, both in deep and shallow water, of a rotifer or wheel animalcule which forms rolling spherical colonies imbedded in a gelatinous medium, each colony consisting of a little cluster of these animalcules arranged in such a manner that their inner ends approach each other in the middle of the mass, while their outer ends, with mouths, cilia, etc., are exposed on the surface. To the naked eye these colonies of rotifers appear like minute grayish specks of floating matter. This species belongs to the genus *Conochilus*, but differs noticeably from the common *C. volvox*. I have thought best, consequently, to describe it as *C. leptopus* (page 256). It was so abundant in the water that a haul of a ring net, a foot across, for fifty strokes of a single pair of oars gave a measured half pint of this form alone.

This colonial rotifer is not to be confounded with the "water bloom," which developed in Yellowstone Lake a little later to an extent very embarrassing to our surface net work. This so-called "bloom" consisted of specks of various algæ growing so freely in the water as to give it a faint tint of dirty green, and washing ashore in quantity along the leeward side of the lake, usually at this season the northern and eastern.

Away from the shore, by far the most common crustacean was *Daphnia pulex*.[*] Although in ordinary situations the males of *Daphnia* are by no means common, in our Yellowstone Lake collections, made in August and September, the males of this variety were many times commoner than the females, making sometimes nearly the whole of a large catch. The few examples of the other sex seen were mostly young, although a female bearing the ephippium occurred occasionally. Next in abundance was the smaller of the *Diaptomi* found in Shoshone Lake, the so-called variety of *D. sicilis*, and with this came somewhat rarely, but still fairly abundant, *D. shoshone* and *D. lintoni*. Several species of *Cyclops* occurred here, only a new one (*C. minnilus*) very frequently, however, and this in small proportion.

Most of these crustaceans ranged in shore as well as in the deeper water of the interior parts of the lake, *Daphnia pulex* falling away in numbers more rapidly in shallow-water collections than *Diaptomus sicilis*. To these inshore species we may add, from our surface-net collections, *Polyphemus pediculus* (sometimes very abundant among the weeds), *Cyclops gyrinus*, and *C. serrulatus* (both rare), *Chydorus sphæricus* (few), *Scapholeberis mucronatus* (few), *Cypris* sp. (only occasional), *Alona*, and the usual miscellaneous drift of shore forms, *Chironomus*, *Allorchestes* (both *dentatus* and *inermis*), *Gammarus*, caseworms, hydrachnids, planarians, *Clepsine*, etc.

Collections made in the lake near enough to the outflow of hot springs to exhibit their influence differed from those made in cold water only in their more scanty character; and where the water was actually warm it commonly contained nothing but the

---

[*] The common and even abundant occurrence of this species in Yellowstone Lake as a form apparently pelagic in its habits (widely contrasted, consequently, with its usual character) was so unexpected and unusual that I hesitated long before assigning this *Daphnia* to the species most abundant in our stagnant pools. Prolonged study of it from various collections in the Park in comparison with those from the waters of Illinois, has finally led me to conclude, however, that this Yellowstone Lake form is not to be specifically distinguished from American examples of *pulex*. In order to furnish material for a more critical comparison than has hitherto been made of the American and European representatives of this species, I append a description, under the varietal name of *pulicaria*, based upon Yellowstone specimens, with figures of both sexes (page 242 and plate XXXVII, fig. 1).

larger crustaceans, insect larvæ especially caseworms, and other alongshore material, together with the dead and empty debris of insect transformation. It was in fact very clear that the frequently observed basking of small fishes in these warm waters was not caused by a greater abundance of their food.

A comparison of the collections made at and beneath the surface, by day and by night, in sunshiny and in cloudy weather, would seem to indicate that the lake variety of *Daphnia pulex* is much more sensitive to sunlight than any other associated form. In collections made at the surface after dark, and in those made in sunny weather below the surface, this was many times the most abundant crustacean; but in similar collections made at the surface in sunshine it was relatively rare, *Diaptomus sicilis* then taking the lead.

Only brief time and scanty opportunity could be had for miscellaneous alongshore work during our visits to Yellowstone Lake. The greater importance, from our point of view, of deep-water work, the stormy weather, and the unfavorable character of the beaches, commonly either covered with bare gravel or packed with large boulders and beaten by surf, made such work unprofitable and difficult. Nearly all our knowledge of this alongshore fauna we owe, in fact, to a day spent in 1891 on Stevenson Island (1¼ miles long by ¼ mile wide), lying 2 miles from the hotel landing, in the north end of the lake. The shores of this island vary from precipitous bluffs on the west to a weedy shallow on the south. The beach is gravelly everywhere, except as boulders thickly cover the bottom and banks, and outside this is a belt of gravelly sand, followed by sandy mud still farther out. Water weeds occur in scanty patches, chiefly *Chara* and algæ; a coating of minute, dirty algæ commonly covers the stones. On the stones are also countless tubes of small *Chironomus* larvæ, almost covering the surface, mostly emptied by the maturing of these insects at the time of our visit. Under the stones are a considerable number of leeches of various sorts (mostly *Nephelis* and *Clepsine*), and an occasional small annelid worm. On them, among the weeds, a small, black, spiral mollusk, *Physa*, may occur by hundreds, and creeping under or over them, or swimming through the water just above the bottom, we found an occasional *Gammarus*. A large species of *Corixa*, several water beetles, *Dytiscidæ*, and a *Perla* larva were all the other insect forms taken here. A small collection of entomostraca from the weeds has not been determined.

Our knowledge of the bottom fauna of Yellowstone Lake is based on the product of eleven dredgings, at depths varying from 15 to 198 feet. Four of these dredgings were in shallow water, 20 feet or less; two were from a medium depth (40 to 80 feet); one ranged from shallow water to deep (20 to 120 feet); one from a medium to considerable depth (40 to 100 feet); and three may be classed as deep dredgings, ranging from 186 to 198 feet. In this deepest water the most abundant inhabitants of the bottom were long and slender annelid worms (*Oligochæta* not yet studied), blood-red larvæ of *Chironomus* of considerable size, and a small bivalve mollusk, *Pisidium*. Several leeches also occurred in our deepest hauls. Nephelis annulata and species of *Clepsine*, a few specimens of *Gammarus*, probably living above the bottom, several small planarians, cloit of living tubes of small *Chironomus* attached to green *Protococcus*, living vegetation—in all probability a drift of *Chara* mud, as the deepest of these hauls was made at not less than 186 feet. In other dredgings, at 100 feet or more, many specimens of *Cypris*, a very few *Daphnia* probably swarming above the bottom, and several worms (nematoids) were added to the above list.

As we worked into shallower water the *Gammari* and the leeches became more abundant, especially *Clepsine*, and at 25 to 50 feet univalve mollusks (*Planorbis exacutus* and *Physa*), *Allorchestes inermis*, small white *Chironomi*, larvæ and pupæ, and caseworms in sand tubes appeared. At 15 and 20 feet, among the weeds, the assemblage of associated animals was *Daphnia*, *Diaptomus*, *Conochilus*, all abundant, many oligochæte worms, leeches and leech capsules, *Physæ*, *Limnææ*, *Allorchestes*, *Gammari*, and cyprids (*Cypris barbatus*, n. sp.), various caseworms, nymphs of *Ephemeridæ*, *Chironomus* larvæ, and larvæ of *Tabanidæ* and *Culicidæ*.

No discussion of the zoölogical resources and relations of Yellowstone Lake with reference to fish-culture would be even approximately complete which did not take account of the animal contents of the streams and other waters connected with it, since these are the principal resorts of the young of the only species of fish the lake now contains and must always be the chief breeding-places of fishes generally. Apart from the river above and below the lake, the most important tributary is Pelican Creek; a peculiar stream for a mountain region in the fact that for 2 or 3 miles of its lower course it is broad, muddy, and comparatively sluggish, more like a bayou than a creek, thick with vegetation, and much frequented by water birds, whose feathers floating on its semi-stagnant surface gave it the appearance of a barnyard pond. Above this stretch, although still bordered by willow-covered and more or less marshy bottoms, it becomes swift and rocky, except where cut across by numerous beaver dams.

The sluggish waters just above its mouth are, as might be expected, rich with small crustaceans and insect larvæ. Amphipod crustaceans were very scarce in this creek, *Gammarus* not occurring in our collections, and *Allorchestes* but once; and among entomostraca, gigantic specimens of *Eurycercus lamellatus* were far the most numerous, making three-fourths of the entire bulk of the product of hauls made in open water and among algæ and other water weeds. With these, in open water, were a new species of *Macrothrix*, a *Diaptomus*, an occasional *Daphnia pulex*, several specimens of *Cyclops*, a few of *Daphnella* and of *Cypris*, and a single *Allorchestes*. Among the algæ, besides the foregoing, several examples of *Simocephalus retulus* were taken, together with *Bosmina*, *Ceriodaphnia*, *Polyphemus*, and *Chydorus sphæricus*. The most abundant insects were, of course, *Chironomi*—larvæ, pupæ, and adults just emerging— and ephemerid larvæ. A few *Corisæ* and caseworms, some small aquatic *Coleoptera*, and a single living *Limnæa* were also noticed.

In the shallower and swifter parts of the stream insect larvæ take the lead, the bulk of the collections consisting of large and small caseworms of various species, most of them attached to stones, larvæ of *Chironomus*, ephemerid and perlid larvæ, large *Corisæ*, and several beetles of small size. The caseworms and ephemerid larvæ were exceptionally common.

On the whole, this stream—which must stand for the present as an example of many others—contained invertebrate forms of animal life in very fair abundance; in the swifter waters the insect larvæ (neuropterous and dipterous) which lurk under stones, and in the more quiet parts entomostraca and insect larvæ of different habit.

A much smaller creek, known on the map as Bridge Creek, and noted among tourists because crossed by a perfect and highly picturesque " natural bridge," has at its mouth a small pond or lagoon which was searched with some care during half a day. This pond is shallow and muddy, but mostly clear in the middle, with a fringe of aquatic vegetation and stout marsh grass growing in the water. The clear water was full of minute spherical masses of algæ, among which were a *Diaptomus* and *Polyphemus*—apparently the usual lake species, but not preserved. In the grass were great numbers of *Gammarus robustus* and a few *Allorchestes dentatus*, and ephemerid larvæ were abundant everywhere. *Chironomus* larvæ were common in the collection and doubtless very abundant in the mud, and a robust *Corethra* larva was occasionally taken. Small water beetles, *Deronectes* and *Haliplus*, caseworms with cases made of fragments of vegetation and others of fine gravel, *Corisa*, a water-skater, *Hygrotrechus*, black *Podurida*, and occasional terrestrial insects were among the other insect elements available for the food of fishes. The mollusks of this little collection were small *Limnæa*, *Physa* (large and small), *Pisidium*, and *Valvata*. Various leeches, the most abundant the common *Nephelis maculata*, *Hydrachnida*, and planarians complete the list thus far made up.

The pond was swarming with young mountain trout (*Salmo mykiss*), a few of which I dissected for a determination of their food. One of these an inch and a half in length had eaten *Chironomus* larvæ and images, chiefly, the remainder of its latest meal consisting of other insect larvæ not in condition to identify, and the entomostracan *Polyphemus pediculus*. A second, an inch and a quarter long, had also fed mostly on *Chironomus* in its various stages of larva, pupa, and imago, but had made about a third of its meal from entomostraca (*Daphnia pulex* and *Polyphemus pediculus*). Another, still smaller, 0.92 of an inch long, taken from the open lake, among the small weeds growing on a flat, muddy rock, had filled itself with *Chironomus* pupæ only, as had still another of the same size. A third specimen from this situation had eaten more larvæ of *Simulium* than of *Chironomus*, and a fourth had also eaten *Simulium* larvæ and another dipterous larva unknown to me.

I may add here that other young trout, in a small, swift rivulet near the Lake Hotel, were feeding continuously, August 9, on floating winged insects, mostly, if not all, *Chironomus* and smaller gnat-like forms.

The large leeches at Bridge Bay (*Nephelis maculata*) betrayed their scavenger habits by collecting in numbers upon a dead fish, which they were evidently feeding from. Two specimens taken elsewhere in this pond proved on dissection to have the alimentary canal nearly empty, one containing only a few fragments of *Gammarus* in the rectum, and the other a single leg of a *Gammarus* in the œsophagus.

As illustrations of the smaller animal life of the river below the lake and in its vicinity, I may report the product of two trips made August 11 and 23, one from a quarter to half a mile below the outlet, and the other to a point about a mile below.

The most fruitful ground at the first locality was a sedgy flat on the left bank and a bed of flat rock covered with algæ and other fine vegetation, with about 6 to 8 inches of water. Other collections were made from the bare sandy bottom, in water 6 inches deep, with moderate current.

On the weedy rocks occurred the large hairy *Capris* described hereafter (p. 211) as *C. barbatus*, and other small blue cyprids not yet studied. The presence of *Chironomus* larvæ, several sorts of caseworms, larvæ of ephemerids, *Hygrotrechus*, *Corisa* larva and adults, various water beetles, specimens of *Gammarus robustus* and *Daphnia pulex*,

Physa (small and large, in quantity), *Pisidium*, many *Limnææ*, of various sizes, *Planorbis*, and small *Amnicolæ*, accounted sufficiently for the fry of the mountain trout abundant among the weeds. The most interesting object here, however, was a small cylindrical brown turbellarian, which has the habit of swimming freely through the water, rolling over steadily from side to side as it swims. Although common here and easily taken, every effort at its preservation failed completely, the specimens going to pieces in spite of the varied use of hot water, corrosive sublimate, cold and hot, osmic acid, Perenyi's fluid, etc. This interesting form seems, notwithstanding, worthy of special mention, and I have drawn up the following brief description, made from field notes, which may serve to identify it to some collector more fortunate than I in his opportunity to study it closely.

Form cylindrical, tapering a little toward both ends, the posterior end blunt-pointed, the anterior flattened in creeping, and broadly rounded. Locomotor surface not specially flattened. When swimming the two ends are similar. Length, when extended, 5 to 9 millimeters, width 1 millimeter. To the naked eye dark reddish or orange, slightly paler before and beneath. Closely examined, the color is in minute, irregular flecks on a yellowish ground, and varies in intensity, of course, according to the extension of the worm. Sometimes the intestine shows through as a darker median shade, and the orange-brown eggs, 0.25 millimeter in diameter, also deepen the color locally. When emptied of these, its color is nearly uniform reddish-brown. The eggs are spherical, conspicuous, in two ovaries, one each side of the abdomen, and, to the number of twenty, may nearly fill the body. A pair of eye-spots placed at a distance from the front of the head about equal to the diameter of the body.

These worms were found in 1891 (September 1), much more abundant than at the above locality, in some clear, gravelly pools filled with filamentous algæ along Soda Butte Creek. They were everywhere thick among the algæ, and could be collected by scores in an hour.

At the lower locality mentioned above, several ephemerid larvæ, specimens of *Gammarus robustus* and of *Allorchestes inermis*, caseworms with cases made of fragments of bark, larvæ of *Simulium*, and some small planarians were found.

Finally I close this preliminary account of our Yellowstone Lake collections by noting the results of a brief examination of the contents of the warm waters along the shores of the hot spring basin of West Bay.

Hauling August 3, 1890, in shallow water only a few feet from shore, at temperatures varying from 70° to 101° F., where the ordinary surface temperature was 62°, we took a great quantity of the rotifer *Conochilus leptopus*, very many examples of *Polyphemus pediculus*, a few *Diaptomi*, and a very few specimens of *Daphnia pulex*. There were probably five times as many examples of *Polyphemus* as of all other entomostraca. The *Diaptomi* were *D. shoshone* (a few) and several *D. sicilis*, and all the other entomostraca were a few each of *Cyclops serrulatus*, *Scapholeberis mucronatus*, and *Chydorus sphæricus*. There were no insects in these collections, living or dead, and the total amount of animal life was much below that of the cold water adjacent. In a spring near shore, with a temperature of 103° F., containing much *Oscillaria* and full of a dark-red alga, there were many holotrichous infusoria and other smaller ciliata, minute flagellata, a fine anguillulid, a small, active planarian, and many examples of a rotifer (*Monostyla*) allied to *M. cornuta*, but apparently new. (See page 256.)

Duck Lake.—As additional material for comparison, the results of a visit to Duck Lake, close beside the West Bay of Yellowstone Lake, may be worthy of present mention. This clear, cold lakelet, about half a mile long and three fourths as broad, lies in a steep, oval hollow of the woods, its shore without beach, too deep for vegetation, and surrounded by a tangle of fallen trees—a secluded woodland pool. The special peculiarity of the little collection brought in from here August 4 consists in the predominance of *Diaptomus lintoni* (the only *Diaptomus* found) over the other entomostraca, and the vast abundance of a shelled rhinsorian (*Difflugia globulosa*) brought up by the dredge at 65 feet and by the towing net sunk to a depth of about 30 feet.

Alongshore a small but miscellaneous collection was made of dragon fly larvae (*Libellulina* and *Agrionina*), larval May flies and *Chironomus*, of several *Amphipoda* (mostly *Allorchestes dentatus*), of various entomostraca, among which were *Simocephalus vetulus*, *Cyclops gyrinus*, and other species of *Cyclops*, *Alona*, etc., and of the common large leech, *Nephelis maculata*. In the dredge, besides the *Difflugia* already mentioned, were several specimens of cyprids (*Candona*), very many *Chironomus* larvae in their tubes, *Cyclops minutus* (a few), a few *Corethra* larvae, *Diaptomus lintoni*, and a small anguillulid worm. The water at some depth was loaded with small pellets of uniform size and similar shape, made up of diatoms, fragments of filamentous and other algae (mostly emptied of chlorophyll), and of other vegetable debris together with grains of sand, all of which had the appearance of being the excrement of the common *Chironomus* larva. So thick was this material that it soon lined the surface net when hauled some 30 feet below the surface. With it came, besides *Difflugia*, a few each of *Corethra* larvae, *Daphnia pulex*, *Diaptomus lintoni* (all females or young), and a single *Gammarus*. A surface haul gave a substantially similar product, with the addition of the entomostracan *Sida crystallina*, not recognized in the adjoining lake.

Lake of the Woods.—Here, as well as anywhere, may be reported the product of a very little work done with the dip net along the margins of a little oval pond a quarter of a mile in length, lying among the hills above Obsidian Cliff, at a height about the same as that of Yellowstone Lake. It has neither outlet nor inlet, and is doubtless fed by springs. It was evidently shallow, although it was not sounded by us, its bottom apparently fathomless mud, and the open water of its center bordered all round by a belt 100 feet wide of pond lilies and the usual accompanying vegetation.

Collections could only be made among the lily pads with a hand net from a log near shore. They were remarkable only for the variety of entomostracan and insect forms and the vast abundance of a *Stentor* which blackened the surface in patches some inches across and covered the lower surfaces of the lily pads as if with a layer of soot. This is allied to *Stentor igneus*, from which it differs, however, by characters to be derived from the description published on page 256. The principal insects taken were ephemerid and *Chironomus* larvae, a few caseworms, a specimen of the water-beetle *Graphoderes fasciaticollis*, many black spring-tails (*Podurida*), and several water spiders. *Sida crystallina* was the most abundant crustacean, but specimens were also taken of *Scapholeberis mucronatus*, *Cyclops*, *Diaptomus*, *Simocephalus vetulus*, and *Acroperus leucocephalus*. A few examples of *Allorchestes dentatus* were also seen, and a fragment of a hairy, bristled worm (*Naidomorpha*).

GARDINER RIVER SYSTEM.

Twin Lakes.—Collections were made August 20 from the upper of two small, closely connected lakes called the Twin Lakes, lying in the boggy trough between the hills beside the main road from Mammoth Hot Springs to the Norris Geyser Basin. This and the two following lakes belong to the Gardiner River system. The Twin Lakes give origin to a small stream known as Obsidian Creek, through which they are connected with a great expanse of swamp and shallow weedy water, known as Beaver Lake. The upper lake is a clear, clean-looking pool, with much marginal vegetation (lily pads and the like), and with boggy banks which drop off suddenly, forming an overhanging grassy margin. Several discolored springs open into the lake, discharging into it water which is said to contain alum in solution.

The dredge brought up from the deepest water found (beginning at 30 feet and ending at 39), a quantity of very soft, streaked, ill-smelling mud, with a little dead vegetation and a very small proportion of animal life. This consisted mostly of *Chironomus* larvæ, partly red, but most of them faded brown, as if discolored by their surroundings. The only other product of the dredge was two specimens of *Gammarus*, a single leech (*Clepsine*), and one *Pisidium*. The water itself, however, was well stocked with animal life, and a haul of a towing net above the bottom, at a depth of 30 feet, at 11 a. m., in bright sunshine, with a stiff breeze blowing, gave a considerable number of *Gammarus*, a very good collection of the characteristic entomostracan of this lake (*Diaptomus lintoni*), and several specimens of *Daphnia* and *Corethra* larvæ. A surface haul under the same conditions gave a few examples of *Daphnia schœdleri*, an occasional *Cyclops*, a single ephemerid larva, and a large quantity of *Diaptomus lintoni*. Alongshore, upon the weedy bottom—an admirable lurking and feeding ground for fish—were the commoner insects (*Notonecta*, *Hygrotrechus*, ephemerid and agrionine larvæ), several specimens of *Gammarus*, a great quantity of the entomostracan *Sida crystallina*, and a few *Simocephalus retulus* and *Chydorus*. Curiously, not a caseworm was taken from this lake—a fact possibly to be explained by the peculiar character of its bottom. A careful search was made from the boat and along the bank for signs of a plant of mountain whitefish made here the preceding year by Mr. Lucas of the U. S. Fish Commission, but no trace of them was found.

Swan Lake.—This lake, a quarter of a mile long by two thirds as wide, is of nearly the same size as the two preceding, but is, perhaps, the shallowest of all (not over 3 feet in depth). It lies on a plateau of the same name, not far beyond Terrace Mountain and beside the main Hot Springs and Geyser Basin road. Its waters are derived from the adjacent mountains to the west, and pass out through Glen Creek into the Gardiner. As it lies in a plain, its immediate surroundings are level. Its bottom is of rock and sandy mud, with *Chara* and other weeds, and a strong growth of rushes inshore.

The collection lists from this little lakelet are unusually full, a fact apparently due chiefly to its geological surroundings. All the waters previously discussed are situated in the Park plateau, and the rocks of their drainage basins are all lava in some form, usually that modification of it known as rhyolite. Swan Lake, on the other hand, is in a cretaceous region, where the geological deposits are largely limestone. We found here, August 29, seemingly in consequence of this fact, an extraordinary abundance of mollusks: large *Limnaea* in the grass at the margin, a small *Planorbis*, *Pisidium*, *Sphaerium*, etc. *Gemmaeus* occurred here under stones, and numbers of *Allorchestes dentatus* were taken in the weeds.

The insects collected were species of *Hydaticus* and *Hydroporus*, *Notonecta* and *Corisa*, caseworms, and *Chironomus* larvæ. A few water spiders (hydrachnids) were taken among the weeds, and on the stones were great numbers of the cocoons of the large leech, *Nephelis maculata*. Young of this leech were numerous, and an occasional adult was seen, with specimens of *Clepsine ornata* and other species of the genus. The most abundant entomostraca were *Daphnia schodleri* and *Diaptomus sicilis*, var., frequent specimens of *D. shoshone* occurring with the latter. That this large species, previously found only in the larger lakes, should replace here in this shallow pond the *Diaptomus lintoni* elsewhere characteristic of shallow water, was another of the surprises of distribution and association of which these mountain lakes are fruitful. The other small *Crustacea* noted were *Cyclops*, *Eurycercus*, *Alona*, and several species of *Cypris*. The occurrence of *Spongilla* here is also worthy of mention.

Gardiner Lakelet.—Similar to the foregoing in geological situation, but smaller, deeper, and surrounded by deep and broken hills, is a little pond formed in the course of a swift and rocky stream to the west of Gardiner River, between Mammoth Hot Springs and the town of Gardiner. It is about 250 yards long by three-fourths as wide, and 20 feet deep in the interior, with its marginal waters filled with a strong growth of rushes and other vegetation. It was full of entomostraca, gammarids, and insects. On the stones were great numbers of the egg capsules of the common large leech, *Nephelis maculata*, and many young of this species were crawling about, but the time for the adults had apparently gone by. The assemblage of entomostraca was again peculiar, the most abundant form being a *Ceriodaphnia*, and the next commonest a medium-sized *Diaptomus* (mostly immature), described on page 253 as a new species under the name of *D. piscinæ*. There were also many specimens of *Cyclops*, an occasional *Daphnia pulex*, and several of *Chydorus*, in a collection obtained by drawing the surface net through the open water. *Chironomus* larvæ were, of course, abundant; and worms of various kinds, small flat planarians, larvæ of *Agrion* and *Dutiscula*, species of *Haliplus*, *Corisa*, and *Notonecta*, larvæ of *Leuctra* and ephemerids, species of *Clepsine* and of *Physa*, were represented in our collections.

Small pools.—Occasional roadside collections made from small standing pools will be of interest chiefly because of the locality and altitude. The ordinary contents of such waters at the time of our visit may be learned from the following lists:

Weedy pool between Norris Basin and Grand Cañon, August 21, 1890:

Chironomus larvæ (a few).	Daphnia schodleri (a few).
Corethra larva (one).	Scapholeberis mucronatus (very many).
Corisa larva (three).	Ceriodaphnia reticulata (many).
Diaptomus lintoni (many).	Chydorus sphaericus (very many).
Cyclops (a few).	Macrothrix sp. (one).
Polyphemus pediculus (a few).	Acroperus leucocephalus (one).
Daphnia pulex (many, with ephippia).	

Weedy beaver pool near Soda Butte Creek, September 1, 1891:

Agabus sp. (two).
Hydroporus sp. (several).
Colymbetes sp. (one).
Dytiscidæ (larvæ).
Corisa sp. (several).
Ephemerid larvæ (a few).

Chironomus larvæ (many).
Phryganeidæ (many, with cylindrical tubes made of cylindrical pieces of vegetation).
Turbellaria (brown cylindrical species. See page 229).

Weedy pond near Soda Butte Creek, September 1, 1891:

Chironomus larvæ (a few).
Corethra larva (one).
Ephemerid larvæ (several).
Cyclops sp. (one).

Simocephalus vetulus (very many).
Ceriodaphnia sp. (very many).
Daphnia pulex (many).
Annelida (one, fragment).

Standing pools left by Soda Butte Creek; covered with algæ; muddy bottom; Sept. 1, 1891:

Hydroporus sp. (one).
Chironomus larvæ (a few).
Ephemerid larvæ (many).
Phryganeidæ (small, one).
Podurid (great quantity on algæ).
Gammarus (one).

Cyclops sp. (one).
Daphnia schœdleri (many).
Simocephalus vetulus (very abundant).
Ceriodaphnia reticulata (many).
Physa sp. (large, two).
Chætogaster sp. (one).

Two circular ponds, each approximately 500 feet across, situated about a mile from Baronette's Bridge and beside Lamar River, were examined in passing, so far as could be done by alongshore work and by wading out with a surface net. In one there was an abundance of vegetation—rushes and a variety of other water weeds—and no appearance of alkaline deposit, the bottom being a film of mud on gravel. In the other there were no rushes, but the water was sufficiently alkaline to have a smooth feeling, and the dead water weeds were whitened as they lay upon the bank.

In the first pond, there was a great quantity of dead shells of a large *Planorbis*, and fewer of a large *Limnæa* around the margin. Vast numbers of *Allorchestes dentata* occurred on the vegetation, and especially in the soft mud of this pond. The entomostraca were nearly all *Copepoda*, of the genus *Diaptomus*, most of them *D. lintoni*. Not a single *Cyclops* was noticed, nor a single *Daphnia*. A few *Ceriodaphnia* occurred, several specimens of *Simocephalus retulus*, a very few of *Chydorus sphæricus*, and, for the rest, a considerable number of hydrachnids, a few *Chironomus* larvæ, and several larvæ of dragon-flies (*Agrion*). With the foregoing were the common large *Corisa* of this region, *Notonecta*, *Physa*, *Deronectes*, and a small hydrophilid larva.

In the alkaline pond near by were a very few mollusks and a moderate number of insects, the latter consisting chiefly of agrionine larvæ and small larvæ of *Chironomus*. The entomostraca were much as before, except that *Diaptomus shoshone* in small numbers mingled with *D. lintoni*.

MADISON RIVER SYSTEM.

Although large collections were made from streams of various sizes whose waters find their way into the Missouri by way of the Madison, the only lakes so connected upon which we worked were Mary Lake and Grebe Lake, the former draining through Nez Percé Creek into the Firehole, and the other giving origin to the main branch of the Gibbon.

Mary Lake.—This little lakelet, situated at a level of 8,200 feet, on the divide between the waters of the Firehole and those of the Yellowstone, is an oval body of water some 5 or 6 acres in extent, clear but shallow, with a fine gravelly beach and very little vegetation. The immediate banks are rather steep, and the country around is densely wooded with pine. The greatest depth at our visit was only 20 feet, and a lagoon-like bay near the lower end was but 5 or 6 feet deep. Both inlet and outlet—the former very small—were dry, but the lake overflows into Nez Percé Creek when the snow goes off in spring.

Our collections from this lake were particularly remarkable for the great number of one of the most beautiful and interesting of our fresh-water entomostraca, the species known as *Holopedium gibberum.* In a haul made with the towing net at the surface, in the shallow water near the outlet, a great quantity of this species was taken, together with a still greater number of *Diaptomus lintoni* and an occasional *Corethra* larva. The sun was shining at the time (11 a. m., August 16) and the water was rippled by a light breeze. Near the bottom, at a depth of 16 feet, *D. lintoni* was the prevailing form, mingled, however, with an almost equal quantity of *Corethra* larvæ and a considerable number of *Daphnia schaefferi.*

The deeper bottom was of sandy mud, which contained a large number of *Chironomus* larvæ in tubes—most of them the common large red species—a great number of the common form of *Pisidium,* and many caseworms with tubes composed of sand grains, several of them bicarinate. The dredge brought up a little *Spongilla,* several of the usual annelids, and *Corethra* larvæ, but no living vegetation. No amphipods were taken from the lake, and no univalve mollusks. An unusual number of aquatic insects occurred alongshore, most of them specimens of *Corixa* of two species, and *Dytiscidæs griseostriatus.* The leeches were, as usual, several species of *Clepsine* and *Nephelis maculata.*

Grebe Lake.—This shallow, muddy lake, connected with the head waters of Gibbon River, is so far secluded within the forest that it has no current name, and is locally almost unknown. We reached it August 27, with saddle and pack animals, from the Grand Cañon Hotel, carrying canvas boat, seines, and our smaller apparatus.

Our collections were made with a surface net in deep and shallow water, with a dredge at a depth of 36 feet (the greatest found), and with the hand net from grass and lily pads near the margin, from the gravelly bottom in shallow water, and from weedy mud—this list exhausting, in fact, all the varieties of situation offered.

The lake lies north and south in greatest length, the outlet leaving the south end and flowing to the south at first. It is in some respects a duplicate of Mary Lake, but is somewhat larger, being half to three-quarters of a mile long and about two-thirds as wide. It is of oval form, with grassy margins, commonly sod to the water's edge, rimmed round with lily pads and other water weeds, and with a bottom of soft, black mud. The banks were somewhat swampy, but the ground was higher to the north and west. Three small streams flow into the lake, one from the northeast and two from the west.

Although so unlike Shoshone Lake, its assemblage of animal forms bore a striking resemblance to that of the larger, clearer lake. The absence of fish, the abundance of *Gammarus* and *Diaptomi*, and the scarcity of *Daphniidæ* are examples. On the other hand, the grassy borders and weedy shallows entertained a much greater abundance and variety of insect forms than the hard and bare margins of either Shoshone or Mary Lake.

In the mud of the bottom were many large red *Chironomus* larvæ, a few specimens of *Gammarus*, and the usual *Pisidium*. The entomostraca were mostly *Diaptomus liatoni*, which replaced in this small lakelet the *D. shoshone* of the other lake; here also we found *Daphnia clathrata*, n. s., the only locality thus far discovered for it. It was not abundant in Grebe Lake, and may have bred primarily in the swamps adjoining. A species of *Cyclops* also occurred here in small numbers, which is described on page 248 as *C. capilliferus*.

Inshore collections were unusually fruitful. *Gammarus* and *Allorchestes* were very abundant along the margin in the weeds and grass, and *Pisidium* especially was extraordinarily common. Here also were agrionine and ephemerid larvæ, caseworms with cases of fine sand, *Corisa*, *Pisidium*, *Physa*, *Chironomus*, and *Spongilla*, and on the mud among grass and algæ were dytiscid and small sialid larvæ, *Physa* and *Pisidium*, *Nephelis* and *Clepsine*, *Allorchestes* and *Gammarus*, *Chironomus* and ephemerid larvæ, larvæ of dragon-flies, and specimens of *Haliplus*.

This lake was an additional illustration of the fact that, in this high mountain region, where aquatic life seems oppressed with unusual difficulties, change in circumstance takes extraordinary effect, so that each lake has its distinct and special zoölogical character.

FLATHEAD RIVER SYSTEM.

The waters of the Flathead region from which collections were made by us were Swan and Flathead lakes and Flathead, Swan, and Cœur d'Alene rivers, and the Jocko at Ravalli. Those from the lakes only can be here discussed.

Flathead Lake.—Although this lake stands in some respects in decided contrast to Yellowstone Lake, these differences tend largely to neutralize each other. Flathead Lake is over 200 miles farther northward than Yellowstone, but the latter is 4,775 feet the higher above the level of the sea. These lakes lie on opposite continental slopes, their waters passing respectively into the Gulf of Mexico and the Pacific Ocean, but neither is more than a few miles from the relatively low continental divide, easily passable by most of the plant and animal forms likely to occur in such waters. Both lakes lie in the course of streams of considerable size, but these streams flow in opposite directions, the inlet of Flathead Lake coming southward from the British Possessions, and its outlet running first to the south and then to the west as Flathead River, a branch of the Columbia, while Yellowstone River, rising about 50 miles above the lake, runs northward more than a degree below it before swinging to the east to join the Missouri. Nevertheless, the headwaters of the two river systems interlace almost inextricably through interlocking mountain valleys along several hundred miles of the main Rocky Mountain range. Both lakes lie among mountains from whose rugged gulches the snow never wholly disappears, and both are bordered by forests broken by park-like openings on the lower slopes; but the geological structure of the surrounding country and the chemical composition of the rocks which form their shores and beds differ widely for the two, and the forests, all pine and fir and other conifers around Yellowstone Lake, are largely deciduous trees about Flathead.

The lakes are similar in size and are both deep enough to give a deep-water character to their interior fauna, but Flathead has much the more uniform shore line and contains—if I may judge from the parts of it which we examined—a larger extent of shallow and weedy water. It is divided, in fact, by a chain of islands stretching across its lower third, into unlike parts; the northern deep and clear, and the southern shallow, and easily stirred up to its clayey bottom by the winds.

Finally, both lakes, like most of this region, are evidently far smaller now than they were in an earlier geological period. The extension of the old Flathead above the present lake is shown by the terraces marking its former shores, which may be traced, one above the other, for a considerable distance above the inlet; while Hayden Valley, the deserted part of the Yellowstone Lake, lies below the lake along its present outlet.

The Flathead is reported by steamboat men and residents to be about 25 miles long by 10 or 12 wide, although the best published map of the region makes it 24 miles long by 17 wide; but as the country about has not yet been surveyed, neither distances nor proportions are precisely known.

The immediate surroundings of this lake are attractive in the extreme. Beside it on the east lies the Mission Range of mountains, beginning to rise almost from the water's edge, and presenting to a near view, along the lower half of the shore, a curiously reg ular series of high, scantily wooded ridges and rocky gulches transverse to the length of the lake. Farther back the peaks of the higher mountains rise bare and steep. This Mission Range diminishes in height northward, and falls away to Swan River, near the northeast part of the lake, but across the river to the east and north the Kootenai Range continues far up along the Flathead. Opposite Mission Range, on the western side of the lake, lies a mass of heights between mountain and hill, rising one above another, mostly wooded, but with occasional park-like openings. Above the lake a level valley several miles wide, partly densely wooded and partly prairie, extends above Kalispell, and to the south lies the naked plateau of the Flathead Reservation.

The principal tributaries are the Flathead, a still, broad river, larger than the Yellowstone at the lake, running from Demersville, most of the way between flat, low banks; the Big Fork or Swan River, a rocky stream, whose course from Swan Lake to the Flathead is an oft-repeated alternation of wild rapids and comparatively quiet reaches; and Dayton Creek on the west, which I did not see. The outlet (Flathead River) flows rapidly away from the lake between bluffy banks which presently become a cañon.

Although this lake lies in a great trough-like valley, the level of much of which is not far above that of the lake itself, there is scarcely any swampy ground in its vicinity, or weedy standing water connected immediately with it or with its tributaries in the vicinity of the lake. The principal breeding grounds of fish, in fact, appear to be upon these streams at a considerable distance from Flathead Lake, so that for most of the species there is a long migration period.

Our systematic work in the lake was all done in and about the northeast bay in the vicinity of the mouth of the Big Fork, and at the lower end near the outlet.

While on this bay we were the guests of Mr. E. L. Harwood, of Demersville, and of the Helena Rod and Gun Club, whose club-house on the bay was our home, while a steam launch belonging to members of this club afforded the only possible means of access with our apparatus to the deeper waters of the lake.

At this locality, where we remained from the 20th to the 22d of September, two dredgings were made, the first beginning at 76 feet and continuing to 125 feet, and the second beginning at 125 feet and continuing to 153 feet. The surface net was hauled from 8 a. m. to 9 p. m., in deep and shallow water, and collections were made with nets and by hand alongshore, among the weeds, from driftwood, and from stones.

Our only temperature observations were made at noon of a bright day (September 22), with a common thermometer only, as no deep-sea thermometer was furnished for this trip. At this time the temperature of the air was 70° F., that of the water at the surface 68°, and that of the mud brought up in the dredge, in a haul commencing at 125 feet and stopping at 153 feet, was 42°.

At the lower end of the lake a heavy storm made work difficult, but we searched thoroughly a rocky flat at the outlet, and collected from the masses of weeds washed up by the waves and from the weedy shallows along the southeast shore.

The open-water collections in Flathead Lake were very similar in general character and in the relative numbers of the principal groups to those in Yellowstone Lake, but the species were all different. In the former lake the so-called *Daphnia pulex* was not once seen, but this species was replaced by a *Daphnia* allied to *hyalina*, and here described as *thorata*. This entomostracan made probably four-fifths to nine-

tenths of the product of every deep water haul with the surface net. *Diaptomus*, the
next commonest form in Yellowstone Lake, was not certainly seen at all in Flathead,
but was replaced by a new variety of *Epischura* (*E. nevadensis*, var. *columbiæ*), which
held practically the same relation to *Daphnia thorata* which *D. similis* held to *D. pulex*
in the other lake. Besides these most abundant pelagic forms we found only occasional
examples of *Leptodora*, *Cyclops*,[*] *Bosmina*, *Scapholeberis*, and *Sida crystallina*, the
last two shore forms which probably would not have been taken very far out. Between the deeper waters and the weedy northern margin of the northeast bay is an
extensive flat of sand, under from 5 to 15 feet of water, and here our tow net hauls
were always remarkably unproductive. Partly, perhaps, because of the barrier offered
by this barren belt of shallow water, the pelagic *Crustacea* did not appear at all in
our alongshore collections as they did in Yellowstone Lake. The assemblage of forms
brought out by the small amount of work inshore which we had time to do, was in
no way remarkable, unless for its deficiencies. *Gammarus* and *Allorchestes dentata*
among the amphipod crustaceans, *Sida*, *Daphnia*, and *Cyclops* among the
entomostraca, species of *Physa*, *Limnæa*, and *Planorbis* among mollusks, and the usual
miscellany of hydrachnids, ephemerid and *Chironomus* larvæ, larvæ and adults of
Dytiscidæ and *Hydrophilidæ*, *Corisa*, planarians, leeches, and annelids—among the
latter, *Pristina leucstris*—were the commoner kinds.

Our first dredging in Flathead Lake was made about 200 yards from land, off the
mouth of a small cove with ideally shores—the first below the Helena Club House—
in water ranging from 70 to 125 feet. The dredge came up, after a haul of about a
quarter of a mile, well filled with soft mud, mostly of slaty color, but somewhat
streaked with reddish brown and mixed with a considerable debris of particles of
dead wood, fragments of dead leaves, cast skins of insect larvæ, and the like.

The greater part of the zoological product of this haul was a mass of the empire
cœnœcium of a species of polyzoan (*Plumatella*, near *arethusa*), and with these came *Chironomus* larvæ, red and pale, a dozen specimens of *Pisidium*, a few cyprids, and a
number of undetermined, slender, pale red, annelid worms, 2 to 3 inches long, and a
millimeter in diameter.

The second dredging was made in the same vicinity, but a little below the preceding and farther out. Beginning about half a mile out from the head of the same
cove, at a depth of 125 feet, we hauled nearly a mile to south and west, taking up the
dredge at a depth of 155 feet, when about three fourths of a mile from the point forming the southern limit of the cove. This haul yielded precisely the same product as
the other—an abundance of the same species in approximately the same ratios.

Neither in variety nor quantity was the animal life of the deeper waters of this
lake, as shown by our work with the dredge and towing net, at all in advance of Yellowstone Lake, with the single exception of the polyzoan of our dredgings, and this
was possibly only a local accident.

The bottom and margins of the southern end of the lake seemed comparatively
barren, the weeds washed ashore containing, in fact, scarcely anything but *Allorchestes dentata*, distorted and cabenated larvæ, and *Corisa*. From the stony bed of
the outlet a considerable supply of masses or arms of several species was obtained. Corisa,

---

[*] Mr. Herrick refers to a species allied to *C. pulchellus* of the Great Lakes, but differing in its more
slender, more acutely curved form and in the armature of some of its legs.

water beetles (*Hydrophilidæ* and *Dytiscidæ*), and perlid and ephemerid larvæ, together with a number of leeches—both *Clepsine* and *Nephelis*—*Physa*, *Limnæa*, *Planorbis*, *Pisidium*, *Gammarus*, *Plumatella*, and a fresh-water sponge (*Spongilla fragilis* Leidy).

Swan Lake.—This lake, visited August 24, lies in the course of the river of the same name, about 8 miles from our camp on Flathead Lake. It is a long and narrow lake—about 12 miles long by 1 to 3 miles wide, according to our guide—lying deeply secluded between two lofty mountain ranges, Mission Range on the west and the Kootenai on the east. Its waters are clear and its bottom is extremely irregular, if one may judge from the soundings made by us at the lower end, for some 2 miles above the outlet. The deepest water (not exceeding 30 feet) lay off the highest points, where the mountains come down to the water's edge, or in pot-holes and winding irregular channels, with weedy bars and banks between. At the upper end, the lake is said to expand to its greatest width and to be surrounded by meadows and marshy flats with water weeds extending a long distance out; and there are similar weedy flats along the shore below, especially at the mouths of creeks. Swan River, above the lake, was also reported to be marshy for some distance up, as is also Spring Creek, emptying near the head of the lake. About half a mile above the outlet were two small islands with gravel beaches and surrounded by shoal water full of rushes and a good growth of other aquatic plants.

The animal species in the deep open water of this lake were precisely the same as those commonest in the larger lake below; namely, *Daphnia thorata*, *Epischura nevadensis*, var. *columbiæ*, taken in numbers, and *Leptodora hyalina*, occurring only occasionally. These pelagic entomostracans were, however, much less abundant in Swan Lake than in the larger, deeper body of water.

The bottom forms were not collected by us, as we had brought no dredge, but the surface net was hauled repeatedly among the weeds in water about 10 feet in depth, and stones, round and small, around the margins of the islands were thoroughly searched. Among the weeds, the commonest entomostraca were *Sida crystallina*, *Eurycercus lamellatus*, and *Cyclops gyrinus*—the latter numerous—and with these occurred a very few specimens of *D. thorata* and of a species of *Alona* undetermined. A few small *Chironomus* and ephemerid larvæ, *Corisa*, *Agrion* larvæ, *Allorchestes dentata*, *Physa*, *Planorbis*, *Hydrachnidæ*, several bristled annelid worms, and a small leech, were also taken here. From the stones along the margin of the island we took great numbers of *Gammarus* and *Cypris*, a very fine sponge occurring in thickish masses on the rocks (some white and some chlorophyl-green), a branching polyzoan (*Plumatella*) clinging to the stones, several species of leeches,—including *Nephelis maculata*, so common in the Park,—planarians, specimens of *Physa*, *Planorbis*, and *Pisidium*, and the usual aquatic insect forms, larval and adult; viz, larval *Ephemeridæ* and *Chironomus* and other small dipterous larvæ, *Corisa*, aquatic *Coleoptera*, *Hydrophilidæ*, *Dytiscidæ*, and several kinds of caseworms. These alongshore collections were, in fact, decidedly larger and more varied than those from Flathead Lake, a fact doubtless to be explained in part by the relatively greater amount of shallow weedy water, and the consequent greater abundance of minute plant and animal life.

DESCRIPTIONS OF SPECIES AND VARIETIES.

CLADOCERA

Daphnia clathrata n. sp.

A species of moderate size, with short, deep head, medium to very long posterior spine, minute pigment speck, and pectinate tarsal claw. In the immature female there is a prominent angle just above the swimming antenna, like that of *D. dentifera*.

In the adult female the head, measured vertically across the rostrum, is twice as deep as its length from the base of the antenna to the middle of the front. It is sharply keeled rather than crested, very broadly rounded, its lower margin very slightly convex or quite straight, and its rostrum well marked in the adult.

The eye is close to the front, the transparent orbit reaching to the margin of the head, of medium dimensions, its antero-posterior diameter contained twice in the space between the eye and the posterior margin of the head. The pigment speck is very minute, placed behind the lower half of the eye and nearer the posterior margin of the head. The fornices are not prominent. Beginning midway between the antenna and the eye, they arch broadly above the base of the former, making an obtuse angle a little beyond the antenna, and continuing as a slight carina backwards and downwards for a little distance on the side of the valves. The ventral margin of the shell is more broadly arched than the dorsal, the latter being, in the immature female, nearly straight from the heart backwards. The valves are conspicuously quadrangularly reticulate, spinose on their lower edges nearly to the beak, and on the upper edge to the vicinity of the heart. The posterior spine is very long, straight, slender, spinose to the tip, contained in average cases not more than twice in the length of head and body without the spine.

The antennae are rather short, about half as long as the distance from the posterior margin of the eye to the base of the posterior spine. The swimming hairs are two-jointed, the basal joint the shorter. The dorsal abdominal processes arise in immediate connection, but are not united at their base. The anal furrow has about a dozen teeth on each side, and the caudal claw has a comb of three or four conspicuous teeth at its base, besides a little group of smaller ones.

Length of an ovigerous female, 1.7 millimeters to the base of the spine; the greatest depth, 0.85.

The male was not seen.

Occasional in Grebe Lake, Yellowstone Park.

Daphnia arcuata, n. sp.

Head helmeted, rounded in front, length one third that of the shell, front concave, beak produced, extending beyond the sensory hairs of the antenna. Fornices beginning above the eye and extending nearly to the middle of the back, not especially produced above the antennae. Eye small, about midway between the mandibles and the front of the head, and about midway between the tip of the beak and the dorsal surface of the head. Pigment speck very small, less than half the diameter of a lens of the eye, and placed midway between the eye and the posterior margin of the head. The latter concave, the beak extending backward and applied against the margin of the shell. Swimming antennae reaching the middle of the shell, their hairs moderately robust, the first segment not longer than the second. No emargination separating head from body, but the dorsal surface very slightly sinuate there. Abdominal processes all distinct, anterior much the longest. Caudal claw with accessory teeth, about six in number, the three proximal the largest. Anal spines about ten. Posterior spine of the shell given off from the middle of the valves, in the adult female a third to a half as long as the valve. Shell moderately coarsely marked in quadrangular areas, the lower and dorsal margins spinulose from the middle backwards. Length 2 millimeters, depth 1 millimeter.

Heart Lake, Yellowstone Park, Wyoming.

Daphnia thorata, n. sp.

This species belongs to the *hyalina* group, and may possibly be entitled only to varietal rank. The distinctness and constancy of its characters, however, in collections made by us in Flathead and Swan lakes in western Montana, and the number of minor points in which it differs from *hyalina*, as most recently described, lead me to distinguish it here as a specific form.

It is oval in outline; the long and slender posterior spine is placed at or a trifle above the middle; the length of the head is about a third that of the valves of the shell excluding the spine, and there is no trace of dorsal emargination between head and body. The head is narrowed toward the base and elongated forward in a way to give it the outline of a high bell jar with a flaring base. Its front is broadly and regularly rounded, its ventral margin usually conspicuously concave and closely like the dorsal, although occasionally the head is straight or convex beneath. The posterior margin is either straight or slightly concave, and the beak stands free from the front margin of the valves, and by its extension downward not only covers the antennæ but reaches clearly beyond the tips of the sensory hairs. The eye is of medium size, placed far back of the front of the head and equidistant from the tip of the beak and the dorsal junction of the head and body. The pigment speck is of moderate size, placed directly behind the eye, and much farther from it than from the posterior margin of the head.

The antennæ are moderately stout, entirely smooth except for inconspicuous transverse rows of minute appressed hairs upon both peduncle and rami, and a row of short, tooth-like spinules at the distal end of each segment. The swimming hairs are rather slender, the second joint commonly decidedly shorter than the first.

Fornices slight, arising above and a little behind the eye and terminating directly behind the antennæ, above the bases of which they project but slightly. The lower margin of each valve is set with the usual spinules almost to the beak, and the dorsal margin is similarly armed for a distance in front of the spine about equal to half the length of the latter. The valves are marked off by fine lines into large quadrate meshes.

The dorsal abdominal processes rise separately, the two anterior, however, in immediate contact at their bases. The first of these is decidedly the longer, but the third process is distinct, although low. The anal setæ are two-jointed, the second joint the shorter. The abdomen is regularly narrowed backwards, and the anal groove is provided with twelve to fifteen teeth on each side, commonly the latter number. The terminal claws are without accessory comb. The intestinal cœca are short, not longer than the diameter of the eye, and extend directly forward.

Length, 2 millimeters to 2.5 millimeters; depth half the length, sometimes a little more. Length of spine somewhat variable, but commonly about equal to the depth of the shell.

Described from females only.

Abundant in Swan and Flathead lakes, Montana.

Daphnia pulex, var. pulicaria, n. var. (Plate XXXVII, Fig. 1.)

Similar, especially in the female, to typical *D. pulex*, to which it is closely related by its more general characters. Body a broad oval, moderately thick, colorless, commonly without dorsal emargination between the head and thorax, although sometimes in the generation of females bearing ephippia there is a broad concavity just above the heart. The lower border of the head is broadly concave and the beak is long and applied against the anterior margin of the shell. Moderately long posterior spine placed above the middle line; caudal claws with two sets of teeth, and with 14 to 17 curved spines at the anal furrow.

The head of the female is small, somewhat depressed, crested, as in *D. pulex*, the crest extending backward to the middle of the dorsum; fornices terminating posteriorly opposite the heart, and extending anteriorly to the eye. The beak projects a little beyond the tips of the sensory hairs; the eye is large, its vertical diameter contained scarcely twice in the distance from the eye to the tip of the beak, placed close to the broadly rounded anterior margin of the head, and provided with many large lenses. Pigment speck of moderate size, midway between the eye and the posterior margin of the head.

Antennae but moderately developed, destitute of scale like appendages like those of *pulex*, but set with inconspicuous transverse rows of rather slender hairs. Swimming hairs moderate and moderately feathered, three-jointed, the third segment very short, but evident. The caecum of the intestine strongly curved, extending at first obliquely downwards towards the middle of the eye, and then turning almost directly upward at an acute angle, terminating midway between the middle of the upper margin of the eye and the front of the base of the antenna. The surface of the valve is marked with quadrate areolations, and the margins, both dorsal and ventral, are provided with backward projecting spines or thorns as far as the middle. The anterior half of both margins smooth. The posterior spine is variable in length, reaching in adult females a fourth the entire length of the head and body.

Dorsal processes of the abdomen distinct, the two anterior contiguous in their origin, not united at their base, the first the longer, smooth, and directed forward, the second hairy, turning backward. Two others in the form of low elevations, the last inconspicuous, but both hairy.

The abdomen is rather broad, the posterior margin broadly rounded, the anterior margin straight; 13 to 17 spines bordering the anal furrow, length regularly increasing from above downwards; the teeth of the caudal claw in two groups of from four to six each, the upper group very much the smaller; the anterior margin of each claw with two distant slight emarginations, as in *D. pulex*.

Length of an adult female 1.9 millimeters without the spine; depth, 1.1 millimeters; spine 0.5 millimeter. Female bearing ephippium, a little deeper (1.2 millimeters).

The male smaller, narrower, with head more depressed, the dorsum especially more nearly straight, and the posterior spine standing higher, continuing the line of the dorsum backwards. The lower margin of head is only slightly concave, the posterior half of it straight. A slight beak is formed just below the sensory antennæ, the latter being attached at a small angular emargination at the posterior angle of the head. From this emargination the posterior margin of the head passes directly upwards in a broad and gentle curve. The eye is very large, placed at the very front of the head. Its longitudinal diameter is contained but once in the head behind the eye. Sensory antenna slightly clavate, slightly expanded at the middle, its length equal to the vertical diameter of the eye. In front of the terminal group of sensory hairs is a long terminal spine, nearly as long as the antenna itself, slightly curved backwards and segmented at the middle. Accessory hair distant from end, but a little below the middle.

Length without spine, 1.4 millimeters; depth 0.9 millimeter; spine, 0.33 millimeter. A single hairy dorsal abdominal process, as in *pulex*.

Yellowstone Lake and other waters of Yellowstone Park.

Daphnia dentifera, n. sp. (Plate XXXVII, Fig. 2.)

This species is broad oval in form, has a long beak and a very large eye, a posterior spine placed high up, and in the male and young female a prominent angle on the dorsal outline between heart and eye.

The head is broadly rounded, with eye close to the front margin. The fornices are short, rising above and behind the eye and extending backwards a little beyond the base of the antennæ, where they form a prominent angle. Thence a slight lateral keel of the valve is continued downwards and backwards a distance about equal to the length of the fornix. The lower margin of the head is broadly concave, the beak produced, projecting as far as the ends of the sensory hairs. The large eye, with numerous lenses, is contained not more than twice in the distance from eye to beak, its diameter a little greater than that of the base of the antenna at its insertion. Pigment speck of moderate size, circular, immediately behind the eye and nearer to that than to the posterior margin of the head.

The head is slightly crested, and the crest, extending backward to the heart, rises over the antennæ in a prominent, nearly rectangular process, still more acute in the young, the tip of which is commonly truncate and bears two or three teeth inclining forward. In the egg-bearing female this process is reduced to a mere obtuse angle, or, in the last generation (that bearing the ephippium), disappears entirely. In young adults this dorsal angle is midway between the eye and the heart, but when fully developed it is on a line drawn from the anterior margin of the valve to the middle of the base of the antenna. The setæ of the antennæ are all two-jointed, the basal joint distinctly the longest. The posterior spine of the carapace is long, slender, and weak, and is commonly contained three or four times in the head and body without the spine.

The margins of the valves are set below and behind with slender thorns, as is also the posterior spine, these thorns extending forward a little distance upon the dorsal margin of the shell. The curved spines bordering the anal furrow are thirteen in number; the caudal claws are without accessory teeth; the surface of the shell is marked with quadrangular areolations.

The first and second dorsal abdominal processes are about equal in length and arise in immediate contact, the anterior turning forward and the posterior backward.

Mature female 1.8 millimeters long by 1 millimeter deep.

The male of this species is smaller than the female; the head is smaller and narrower, the form is more nearly elliptical, and the dorsal angle is as prominent as in the young female, and commonly bears obscure teeth at the tip. The head is subquadrate, with rounded angles. The very large eye is at the extreme front of the head, its diameter greater than the distance between the eye and the posterior margin of the head. Below, the head is straight; the anterior antennæ are not especially prominent and the terminal spine is inconspicuous. The posterior spine is like that of the female, long and slender and dorsally placed. The abdomen is without dorsal process. Mature specimen 1 millimeter long by 0.5 millimeter deep.

Closely allied to *D. dentata* Mathle, with which my friend Professor Birge considers it possibly identical. It differs, however, particularly in the form of the head, the beak of which is much more produced backward in *dentifera* than in *dentata*; in the somewhat larger eye, especially of the males; in the different form and position of the dorsal angle, and in its evanescent character in the female adult; in the greater length and slenderness of the posterior spine; and, notably in the male, in the different armature of the anterior antenna. *Dentifera* is also without the third joint of the swimming hairs of the antenna.

Pool near Shoshone Lake, Yellowstone Park.

OSTRACODA.

Cypris barbatus, n. sp. (Plate XXXVII, Figs. 2 and 3, and Plate XXXVIII.)

An extremely large, very hairy, oblong *Cypris*, with rounded ends and dorsal and ventral margins nearly parallel. Length, 4 millimeters; width, 1.6 millimeters; depth, 2 millimeters. A very little deepest at hind end of hinge margin. (Depth across eye, 95 per cent of greatest depth.)

Dorsal margin about straight for a great part of its length, the ventral margin very slightly emarginate or sinuate at its anterior third. The anterior end broadly and smoothly rounded, more obliquely above than below, the posterior somewhat obliquely rounded, the ventral margin being thus nearly half as long again as the dorsal. Seen from above the shape is symmetrical, a slender oval, a little more flattened at the sides behind than before; thickest, consequently, before the middle.

Color a dirty yellowish brown in alcohol, with a reddish brown patch on either side above and behind the middle. Surface of valves opaque, very minutely rough-ened, and well covered with conspicuous hairs, which give this *Cypris* a decidedly hairy appearance to the naked eye. Hairs longest before and behind and becoming generally towards the margin, where they project as a fringe, the most prominent part of which is a row of hairs borne on slender conical tubercles within the margin of the valves. The valves are equal and the shell curls full but not plump.

Anterior antenna with the basal segment obliquely channeled, partially dividing it into two, the distal part of which bears a single bristle on its superior surface, and two long, more slender ones, springing together from the tip of the ventral surface. A short, subquadrate second segment bears a single seta, about as long as the segment, on the dorsal surface, near the tip. From the distal end of the following segment spring two long, slightly plumose setæ, one dorsal, one ventral, the former much the longer. The fourth segment bears at its tip four long setæ, two of which arise from the ventral angle and two from the outer dorsal. The following segment is similarly armed, and the distal extremities of the sixth and seventh are densely set with long plumose setæ forming a stout fascicle, which extends beyond the end of the antenna a distance equal to the length of the antenna itself.

The terminal segment of the palp of the first maxilla is a little more than a fourth the length of the basal, the latter with one subterminal bristle without, and several terminal ones. Tip of last segment with two stout, curved, claw-like setæ, and four or five smaller, softer ones. Outer lobe of maxilla proper reaching to tip of first segment of palp, nearly equaling it in diameter, also with two curved claws, shorter but much stouter than those above mentioned, three-fourths as long as the lobe itself. Besides these, two smaller setæ and three or more subterminal ones, two of which are smooth, like the terminal group, and one strongly plumose. A single plumose seta also springs from near the base of the concave surface of this lobe. The second and third lobes similarly armed at tip, but with a larger number of curved setæ, all of which are soft. Two of these, on the short inner lobe, are much longer and stouter than the others, and project directly backward. The base of this lobe bears two plumose setæ about as long as those just mentioned. The length of the inner lobe is half that of the outer, the middle one being intermediate.

The second maxilla with about twelve terminal setæ, which diminish in length inward, most of them slightly plumose, and two long slender setæ, one springing from the middle of the inner margin and the other from the base. Palp thick, slender ovate, twice as long as the masticatory lobe, fringed with a soft silky pile, and bearing three more or less plumose setæ at its tip, the middle one of which is the longest. Branchial lobe very small, semicircular, with three fully developed plumose setæ nearly as long as the palp, and two much shorter ones, one delicate and smooth, the other stout and plumose.

The basal segment of the second antenna trigonal, with one moderately long hair beneath, and two of similar length springing together from the inner side of the apex. The second segment subcylindrical, with two hairs diverging from the middle of the outer side of the apex, the under one of which is very short and weak, about as long as the third segment is wide, while its companion reaches about to the tip of that segment. On the inside of the tip of the second segment is another hair, similar to the above, and of about the same length. The third segment bears, at the union of its basal with its middle third, on the under side, set beyond a slight tooth-like projection, a jointed olfactory club, whose length is about two-thirds the diameter of the segment. Otherwise this segment bears no hairs except at the tip, where, upon its inferior angle, is one long, stout hair, reaching beyond the tip of the last joint, and upon its inner surface a fascicle of five plumose hairs, the four longer of which are curved and parallel, while the fifth is short and straight. The third segment is slightly longer than the second and about two-thirds as thick. The fourth segment is three-fourths the length of the third and about two-thirds its diameter, slightly enlarged at the middle, where it bears, on the under side, a group of three long hairs, and upon the upper side two shorter ones. At the tip of this segment are a group of three long plumose hairs and a stout, curved, concave, acute claw, nearly three times the length of the last segment, doubly dentate on both edges. At tip of last segment the usual strong, curved, bidentate claws, five in number, three of equal length, as long as the two last segments of the antenna, and two others about half that length.

Mandible with a row of six dark corneous teeth, more or less bifid, the series continued in an irregular cluster of tooth-like spines, and terminating in two highly plumose setæ. The series of teeth with numerous accessory smaller teeth and spines, and two transparent lamellæ—slender, but as long as the teeth themselves—inserted between the first and second and the second and third series, respectively. The latter lamella is recurved and serrate on its concave edge. Basal segment of the palp longest, the third next, second and fourth subequal in length, the second as broad as the first. The latter bears at its posterior tip three plumose setæ of unequal length, in a cluster, and a fourth larger, stouter, decurved articulate one, inserted on the outer side of the tip of the segment. The second segment has in front a group of three slender setæ inserted a little behind the tip; and opposite to them upon a stout tuberosity another group of three long equal setæ, to which a fourth stands in the same relation as on the preceding segment. On the third segment is a group of five setæ similar to those on the anterior margin of the segment preceding; and, in addition, a circlet of six, attached around the posterior and inner margin of the end of this segment. At the tip of the palp are three curved claws, averaging as long as the two preceding segments together, with some slender setæ intermixed. The so-called branchial appendage is about as long as the basal segment of the palp, and bears four stout plumose setæ with a small accessory seta in front.

First leg with basal segment columnar, distal portion partially separated, without hair or bristle. Second segment cylindrical, its surfaces smooth except for numerous transverse rows of exceedingly fine short setæ, present also on the two succeeding segments of this leg. A stout bristle at anterior distal angle. Third and fourth segments nearly equal; the third, however, somewhat the longer; together slightly longer than the second, the length of each about twice its transverse diameter. The third with a single apical hair at the anterior angle, and the fourth with but two, one of which is as long as the segment itself, and the second about half that length. Terminal segment with a very long, slender, symmetrically curved, regularly tapering claw, with two short soft setæ springing from its base. The entire claw somewhat longer than the last three segments conjointly.

Caudal rami long and slender, slightly sinuate, the transverse diameter of each not more than one-twentieth its length; the basal fifth, however, considerably thickened. Rami smooth, except posteriorly, where the margin is closely set with stout, short spines, lengthening toward the distal end of the ramus. Terminal claw slightly curved at tip, contained two and a half times in the length of its ramus. Subterminal claw nearly two-thirds the length of the terminal, also slightly curved. Claw-like seta almost immediately above the latter, more slender, but two-thirds its length. Besides the above, a short slender seta springs from in front of the base of the terminal claw.

The first and last segments of the second pair of legs subequal, each two-thirds the length of the second; basal segment straight, its length five times its width, with three slender setæ, one borne upon the middle of its exterior side, and two near the apex, opposite each other. Second segment slightly curved, with a single slender seta near the apex, on its outer margin. Third segment with two terminal setæ, one nearly straight, and claw-like, about three-fourths the length of its segment, and the other curved and blunt.

This species may be the same as *C. grandis* Chambers[*], which it certainly seems to resemble closely, but from which it differs, if I may judge from the published brief description and rude figures, in color, surface, form, arrangement of antennal setæ, and other minor details.

Yellowstone River, Yellowstone Park, Wyoming.

[*] "New Entomostraca from Colorado," in Bull. U.S. Geol. and Geogr. Surv., vol. III, No. 1, p. 151.

COPEPODA.

Cyclops minnilus, n. sp.

A small slender species, with seventeen-jointed antennæ, with narrow and loosely articulated cephalothorax and salient thoracic angles, slender abdomen, long and narrow furca, and but two well-developed caudal setæ for each ramus. The antennæ reach to the posterior margin of the second distinct segment, and are of very nearly the length of the abdomen (including furca, but excluding the caudal setæ). The greatest width of the thorax is contained two and one-third times in its length, and the furca is very nearly half the length of the remainder of the abdomen. The diameter of a ramus is about one-seventh its length.

The rudimentary inner caudal seta is a trifle longer than the outer, and about a third the length of the ramus; the longest seta as long as abdomen and furca; the next in length less than half the longest.

The last segments of the thoracic legs are armed as follows:

First pair: outer ramus, one spine and two setæ at tip, two setæ within, and one seta without; inner ramus, one spine and one seta at tip, three setæ within, and one seta without.

Second pair: outer ramus, one spine and one seta at tip, three setæ within, and two spines without; inner ramus, one spine and one seta at tip, three setæ within, and one seta without.

Third pair: outer ramus, two spines and one seta at tip (second spine twice as long as first), three setæ within, and one spine without; inner ramus, one spine and one seta at tip, three setæ within, and one seta without.

Fourth pair: outer ramus, two spines and one seta at tip (second spine twice as long as first), three setæ within, and one spine without; inner ramus, two spines at tip (one twice as long as the other), two setæ within, and one seta without.

Rudimentary legs of fifth pair distinctly articulated, basal article with a long seta at its outer distal angle, and second article with two setæ at its blunt tip, the outer the longer.

Duck Lake.

Cyclops serratus, n. sp.

A very long, narrow, loosely articulated species, with strikingly salient thoracic angles; cephalothorax broadest far forward and lobed in front, between the seventeen-jointed antennæ.

Abdomen long and slender, with very long and narrow caudal rami, and but two developed caudal setæ to each ramus. The first segment is but little longer than wide (eight to seven), is broadest across the middle, and excavate in front at the base of each antenna, leaving a thick, median, projecting lobe. The second segment is nearly a fourth as long as the first, and but little narrower, broadest across its posterior angles, which, though blunt, are so strongly salient that the lateral margins are decidedly sinuate. The third segment is as long as the second, but narrower, and with its sides more nearly parallel. The fourth and fifth segments are progressively shorter and narrower, the latter being trapezoidal, as seen from above, and separated from the first abdominal segment by a deep acute emargination.

The abdominal segments are as long as the cephalothoracic segments two to five taken together, and the furca is as long as the last three segments. The first segment of the abdomen is broadest in front, where its width is nearly as great as its length. The second is as broad as long, the third and fourth equal, the fifth a little shorter, the last with a row of fine spinules around the base of the rami.

The width of each ramus is contained nearly eight times in its length. Besides the lateral spine—situated a little before the posterior third of the ramus—there is a cluster of two or three minute spines at its anterior fourth. The outer and inner terminal setæ are reduced to short subequal spines about twice as long as the ramus is wide. The other setæ are slender, plumose, the inner nearly twice as long as the outer of each pair.

The antennæ are rather stout and short, seventeen jointed, reaching to the end of the second segment. They are without special structures or appendages.

The last segments of the thoracic legs are armed as follows:

First pair: outer ramus, one spine and two setæ at tip, two setæ within, and one spine without; inner ramus, one spine and one seta at tip, three setæ within, and one seta without.

Second pair: outer ramus, one spine and one seta at tip, three setæ within, and two spines without; inner ramus, one spine and one seta at tip, three setæ within, and one seta without.

Third pair: outer ramus, two spines and one seta at tip (one spine twice as long as the other), three setæ within, and one spine without; inner ramus, one spine and one seta at tip, three setæ within, and one seta without.

Fourth pair: outer ramus, two spines and one seta at tip (one spine double the length of the other), three setæ within, and one spine without; inner ramus, two spines at tip (one double the length of the other), two setæ within, and one seta without.

The fifth pair is two jointed, the basal joint broad, quadrate, with a seta at its outer angle; the second cylindrical, with one long and one short seta at tip.

Length, without setæ, 1.34 millimeters.

Described from females only.

Cyclops capilliferus, n. sp. (Plate XL, Figs. 14–17, and Plate XLI, Fig. 18.)

This is a symmetrical, compact *Cyclops*, with the cephalothorax closely articulated, widest at the middle, and the sides regularly convex, with the abdomen narrow and slender, with three well-developed caudal setæ, and sixteen jointed antennæ bearing several very long setæ.

The abdomen, with caudal furca, is contained a little less than twice in the cephalothorax, and the breadth of the latter is just half its length. First segment very long, five times the length of the second; second and third equal; the fourth very short, on the median line semicircularly excavate behind. The abdominal segments in the female diminish regularly in length from first to last. The caudal rami are twice the length of the last segment and one fourth as broad as long. The lateral seta is placed a trifle beyond the middle of the ramus; the outer terminal seta is a short naked spine; the other three are well developed and plumose. The inner and outer of these three are of nearly equal length, the latter a little the longer, the middle one much the longest one of the group.

Antenna moderate, reaching about to the middle of the second segment of the cephalothorax. Sixteen-jointed in all adult females, and further especially distinguished by the presence of very long flexible setæ upon the first, third, tenth, and fourteenth segments. Terminal setæ likewise very long. The seta borne by the first segment extends to the twelfth; that of the third reaches to the fourteenth; that upon the tenth segment extends to the tip of the antenna, and that upon the fourteenth far beyond it. All these foregoing setæ are borne upon the anterior terminal angles of their segments with the exception of that of the fourteenth, which is borne upon the posterior angle.

The first segment is as long as the two following, and very nearly twice as long as wide. The second is very short, its length one-fourth its width, and the length and width of the third are equal. Of the three terminal segments the penultimate is longest, being twice as long as broad; the antepenult two-thirds the length of the following; the last is about as wide as long.

The last segments of the thoracic legs are armed as follows:

First pair: outer ramus, one spine and two setæ at tip, two spines without, and three setæ within; inner ramus, two setæ at tip, one within, and two without.

Second pair: outer ramus, one spine and two setæ at tip (the inner of the latter slender, the outer thick), four setæ within, and two spines without; inner ramus, two setæ at tip, one within, and three without.

Third pair: outer ramus, two setæ at tip (the outer one stout, short, and spine-like), four setæ within, and two spines without; inner ramus, two setæ at tip, three within, and one without.

Fourth pair: outer ramus, two setæ at tip, four setæ within, two spines without; inner ramus, two setæ at tip, two within, and one without.

The fifth pair are two-jointed, the terminal joint with one long and one short seta at tip; the basal joint with one long seta without.

Length, without setæ, 1.2 millimeters.

Grebe Lake, Yellowstone Park.

Cyclops thomasi Forbes. (Plate XXXIX, and Plate XL, Fig. 13.)

Cyclops thomasi Forbes, Amer. Nat., XVI, Aug. (1882), p. 649; Cragin, Trans. Kan. Acad. Sci., VIII, 1881-82, p. 68 (1883); Herrick, Final Report, p. 153 (1884); Underwood, Bull. Ill. State Lab. Nat. Hist., II, 1886, p. 332; Forbes, Rept. U. S. F. C., 1887, p. 707 (1891).

A long and slender species, with seventeen-jointed antennæ, oval cephalothorax, somewhat closely articulated, slender abdomen, very long and slender caudal rami, and two developed setæ to each ramus, the longer of which is about twice as long as the shorter.

The cephalothorax is widest at about the middle, its greatest width a little more than half its length. Posterior angles not prominent or produced, except those of the last segment, which are slightly produced outwards. Sides of the first segment subparallel, rounding slightly towards the front, the segment itself twice as long as the other segments combined; the second segment shorter than the third, but longer than the fourth; the fifth reduced to a narrow linear band, as seen from above, the extremities of which project a little beyond the lateral outline.

Abdomen, with furca, a little shorter than the cephalothorax, its greatest width one-fourth of its length, including furca. First segment in the female as long as all the others together, broadest in front, its lateral outlines emarginate behind the anterior angle. Posterior margin of last segment serrate beneath and at sides; those of other abdominal segments smooth. Furca as long as the last three segments, the width of the rami about one-seventh of their length. The inner of the two longer setæ as long as the entire abdomen, the outer of the two half that length. The extreme outermost of the terminal setæ two thirds the length of the inner; that is, about one fourth the length of the caudal ramus. Rami slightly curved outwards, with one large spine and a few small ones a little beyond the middle of the outer surface, and a vertical comb of small spines at one fourth the distance from the proximal end.

Antennæ of the female moderately robust, reaching about to the middle of the third segment, without special accessory structures or appendages, the three terminal segments gradually increasing in length, the antepenultimate two fifths the length of the last. The two segments preceding the former, taken together, shorter than the last segment and about equaling the penultimate.

First pair of legs: outer ramus, two setæ at tip, two spines without, and two setæ within; inner ramus, one spine and one seta at tip, one seta without, and three setæ within.

Second pair of legs: outer ramus, one spine and one seta at tip, two spines without, three setæ within; inner ramus, one spine and one seta at tip, one seta without, and three setæ within.

Third pair of legs: outer ramus, one spine and one seta at tip, two spines without, and three setæ within; inner ramus, one spine and one seta at tip, one seta without, and three spines within.

Fourth pair of legs: outer ramus, one spine and one seta at tip, two spines without, and three setæ within; inner ramus, two spines at tip, one seta without, and two setæ within.

The outer ramus of the first leg is so foreshortened that the distal outer seta seems to be placed at the tip of the segment, but the usual tooth marking the lateral distal angle of the segment stands between this point and the seta next within, thus showing that the spine should be counted as lateral.

The terminal spines of the inner ramus of the fourth pair are unequal, the inner one a little more than half the outer.

Fifth pair of legs of two segments, the basal segment about as long as broad, with a strong plumose spine from the outer angle, the terminal segment cylindrical, twice as long as broad, with two terminal setæ, the outer of which is as long as the seta of the preceding segment, and the inner a little more than half that length.

Total length, without setæ, 1.33 millimeters; greatest depth a trifle less than one-third the length of the cephalothorax.

The common *Cyclops* of Yellowstone Lake, occurring also in various other waters of that region. This well-marked and constant species has a range at least from the Rocky Mountains to the Atlantic region, being, according to Prof. Cragin, a common species in the water supply of Boston. It is also the usual *Cyclops* of the Great Lakes.

The original description was inaccurate in two particulars: the outer distal spine of the outer ramus of the first leg was called terminal, and, by typographical error, the terminal point of the outer ramus of the second, third, and fourth legs were said to have two setæ within, instead of three.

Diaptomus shoshone, n. sp. (Plate XLII, Figs. 23-25.)

A very large and robust species. Thorax broadest in front, across the maxillæ, tapering gradually, with little convexity, to the posterior third. In the female the angle of the last segment is bifid, both projecting points being minutely spinose at tip. The first segment of the abdomen is laterally expanded; the expansion of the left side with a minute spine at the apex behind; that on the right produced at the same point into a small, prominent, rounded tubercle, 0.03 millimeter in length, about as broad as long, making this first segment somewhat unsymmetrical. This is not merely a modified cuticular appendage, but is penetrated by the hypodermis. Egg mass very large, obovate (narrowest forward).

Right antenna of male robust, the last two joints without special appendages, antepenultimate with a very long inarticulate process at its outer apex, extending beyond the tip of the penultimate, and to the middle of the last segment. The margins of this process are smooth, but it is broad and emarginate at the tip.

The fifth pair of legs of the male resemble the corresponding appendages of *Diaptomus stagnalis*, but differ notably in detail. The left ramus of the right leg is borne at the inner terminal angle of the second joint; is longer than the joint following; is armed at the apex with a few small acute spines; and bears upon its outer margin, near the tip, a broad fascicle of delicate hairs. The basal joint of the outer ramus is two-thirds the length of the second joint of the peduncle, and without hairs or spines of any description. The second joint of this ramus is about equal in length to the second joint of the peduncle, and bears at its outer margin, close to the tip, the usual stout seta, which is two-thirds as long as the joint to which it is attached. The terminal claw is not regularly curved, but is nearly straight for the basal three-fourths. The left leg is biramose, the inner ramus straight, slender, extending about to the middle of the second joint of the outer, and armed at its tip. The second joint of this ramus is as long as the first, if measured from the tip of the apical spine. This spine, seen from behind, is stout, conical, rather blunt, and has opposed to it within, projecting from the inner angle of the segment, a stout, curved seta, slightly plumose on its distal half. Between these, but more closely applied to the outer spine, is a hemispherical cushion-like elevation, set with small, short spinules. On the basal half of the inner margin of this terminal segment is also a much larger hemispherical cushion, but with longer and more slender hairs, while the terminal half of the inner margin of the segment preceding is also moderately inflated and covered with delicate hairs.

The antennæ of the female are 25-jointed, as usual, and reach to the base of the abdomen. The legs of the fifth pair closely resemble those of *stagnalis*, but have the terminal setæ of the inner rami much less developed. This ramus is a little shorter than the basal joint of the outer ramus, and of about half its diameter. It bears at its tip two stout setæ equaling the ramus itself in length, plumose under a high power, and has, in addition, at its inner tip and on the inner margin adjacent, a patch of delicate hairs and spines. The second joint of the outer ramus is as long as the first, if measured to the tip of its terminal claw. The latter is nearly straight, very slightly recurved. This joint bears a single spine at its outer distal angle, just within which is the rudiment of the third segment of the ramus, which bears two spines similar to the above, the inner of which is the longer, the outer itself being longer than the adjacent spine of the second joint. Adults of both sexes are blood-red throughout, except the egg sac of the female, which is purple.

Dimensions of female. Length to tip of caudal setæ, 3.1 millimeters; abdomen, with setæ, 1.16 millimeters; without, 0.67 millimeter; thorax, 1.95 millimeters in length; depth, 0.725 millimeter; width, 1 millimeter.

Male averaging scarcely smaller, but somewhat differently proportioned. Thorax, 1.85 millimeters in length; depth, 0.58 millimeter; width, 0.08 millimeter; abdomen, without setæ, 0.745; with setæ, 1.35 millimeters.

Especially abundant in Shoshone Lake, but occurring in other lakes and even in pools of some size in Yellowstone Park.

Diaptomus lintoni, n. sp. (Plate XII, Figs. 25-28.)

A large red species occurring commonly with *D. shoshone*, but distinguishable from it at a glance by its different shape, its longer antenna, its smaller size, and by characters derived from the right antenna and the fifth foot of the male. The thorax is symmetrically elliptical in shape, broadest at the middle. The posterior angles are not produced or bifid, but are each armed with a minute spine. The first segment of the abdomen of the female is not especially produced, but bears at its broadest part a minute spine on each side. The abdomen itself is very short, its length contained about three and one third times in that of the cephalothorax. The antenna of the female is long and slender, 25-jointed, reaching a little beyond the tip of the abdomen.

The fifth pair of legs in this sex is similar to those of *D. shoshone*, but much smaller. The inner ramus is not jointed. It is longer than the basal joint of the outer ramus, bears two stout plumose setæ at its tip, somewhat shorter than the ramus itself, and has likewise at its inner tip a patch of small spines or fine hairs. The second segment of the outer ramus with its terminal claw is two thirds as long again as the preceding segment, the breadth of the latter two thirds its length. The third joint is indicated by a single long stout seta and one or two smaller ones.

In the male the geniculate antenna is relatively rather slender, its last two joints without special appendages, its penultimate with a slender transparent apical process, reaching about to the middle of the succeeding segment, acute at tip, but neither serrate nor emarginate. Fifth pair of legs in the male usually without internal ramus to the right leg, but this ramus sometimes represented by a small rudiment. The limb is usually slender and its terminal claw short. The basal segment of the outer ramus is nearly as long as the adjacent segment of the pedicel, and the slender second segment of this ramus is fully as long. Long lateral spines borne near the tip of this segment. The terminal claw is about two thirds as long as the segment, is somewhat abruptly angulated near its base and slightly recurved at the tip. The inner ramus of the left leg is very stout and long, reaching almost to the tip of the outer ramus, is slightly curved outwards and has the apex minutely hairy. The basal segment of the outer ramus is thick, two thirds as broad as long, somewhat inflated within, where it extends downward beyond the articulation with the second segment as a rounded expansion covered with extremely fine hairs. Second segment of this ramus longer than first, but only half as wide, bearing at its tip, within, a rather small, obliquely projecting cushion covered with cilia, and with two stout terminal spines, one short, blunt, straight, and smooth, the other curved and plumose, its length about half that of the segment to which it is attached.

The total length of this species is about 2.5 millimeters, excluding caudal setæ; depth, 0.42 millimeter.

This species is closely related to *D. stagnalis* Forbes, from which it differs conspicuously by its smaller size, more symmetrical cephalothorax, without prominent or bifid angles, and longer and more slender antennæ, with longer and more slender appendage to the antepenultimate segment.

In the fifth legs of the female this species differs from *stagnalis* especially with respect to the inner ramus, which is larger and longer than in the other, lacks the characteristic segmentation of *stagnalis*, and bears at its tip shorter and broader setæ. In the male the terminal claw of the outer ramus of the right fifth leg is much more slender than in *stagnalis*, and the inner ramus is much less developed. The left leg of this pair is different in a number of details, especially in the length and strength of the inner ramus and the length and dissimilarity of the setæ at the end of the outer.

Common in lakes and pools of Yellowstone Park.

Named for my friend, and companion on the trip of 1890, Prof. Edwin Linton, of Washington and Jefferson College, Pennsylvania.

Diaptomus piscinæ, n. sp. (Plate XLI, Fig. 22.)

A species of medium size and symmetrical proportions, antennæ reaching to the tip of the abdomen, cephalothorax broadest about the middle, with four distinct sutures, the posterior lateral angles not produced, but armed with two distal spines.

The right antenna of the male is without appendage to the antepenultimate joint, and the fifth pair of legs of the same sex has the inner ramus well developed on both the right and left sides. The usual length is 1.75 millimeters, the transverse diameter 0.45 millimeter; the abdomen, with furca, is a little more than one-third the length of the cephalothorax.

The fifth pair of legs of the female is without especially marked characters, except that the inner ramus, which reaches to the tip of the principal segment of the outer, is provided with two long, stout, equal setæ more than half as long as the ramus itself. The third joint of the outer ramus is aborted, and bears two short, stout spines, and the joint preceding bears a slender spine outside the base of the last. The terminal claw of this joint is simple and nearly straight, viewed in the usual position.

In the male the fifth pair of legs has a considerable resemblance to the corresponding appendages of *D. leptopus*, from which, however, this species differs by its more slender form and by the absence of the antennal hook. The peduncle of the left leg is quadrate and equal in length to the basal segment of the outer ramus, but is nearly twice as wide. The sides of this latter segment are parallel, the inner terminal angle is broadly rounded and minutely ciliate, and to the outer terminal angle is attached the second segment of the ramus. This segment is a trifle shorter than the preceding and less than half as wide, and bears at its tip a stout, blunt, conical spine, whose length is equal to that of the diameter of the ramus, and within this a long flexible hair as long as the ramus itself. The inner ramus of this leg is very long, reaching beyond the middle of the terminal joint of the outer ramus. It is slightly concave towards this ramus and terminates with a broadly rounded or subtruncate, thickly ciliate end, forming an acute outer angle and an obtuse inner one. Seen at right angles to this view, the tip is simply obtusely pointed.

The right leg of the male is without remarkable distinguishing characters. Basal joint of the outer ramus about two-thirds as long as the peduncle and nearly as wide; second joint slightly longer than the peduncle, equal to the first in width; and the terminal claw sinuate or irregularly curved. The stout seta on the outer margin of the second segment of this ramus is borne at about a quarter the length of the segment from the distal end, and is approximately half as long as the segment to which it is attached. The inner ramus is a little longer than the basal joint of the outer. It is not dilated or otherwise modified, but terminates bluntly, bearing at the tip a covering of long cilia.

The right antenna of the male is without notable distinctive characters. The antepenultimate segment is as long as the two following taken together; the fourth from the tip bears two long sword-like spines at its margin, both attached to its basal fourth; the expanded segments are well armed with conical spines, straight and curved, but without hooks.

Small lakelet near Gardner, Montana.

Epischura nevadensis, var. columbiæ, n. var. (Plate XII, figs. 19-21.)

It is with pleasure that I report here the occurrence of another form of this interesting genus of North American entomostraca, the fourth or fifth thus far discovered. The first species described, *E. lacustris*, has been found in the Great Lakes, in the smaller lakes of Wisconsin and Minnesota, and at Portland, Oregon; the second, *E. fluviatilis* Herrick, has been seen only by the original describer of the species, by whom it is said to occur in Mulberry Creek, Cushman County, Alabama.* *Epischura nordenskioldii* Lillj., is from Newfoundland, and *E. nevadensis* from lakes Kenn and Tahoe, the former in California, the latter partly in that State and partly in Nevada. The present form occurs in Swan and Flathead lakes, in northwestern Montana, where it was the most abundant copepod in the open water.

The absence of all representatives of this genus from the lakes of Yellowstone Park, evidently adapted to them, hints strongly at a limit of altitude to their distribution. The highest locality from which any species has been reported is Lake Tahoe, said to be 6,250 feet above the sea; while the lowest lake of suitable size in Yellowstone Park from which our collections were made was 1,200 feet higher than this. This topographical difference does not measure the biological difference, however, as the lower location is also more than five degrees south of the Yellowstone lakes.

Disregarding the doubtful *fluviatilis*, the species of *Epischura* are, so far as known, of north temperate range in North America. The form least modified, both in abdomen and fifth legs, is the Newfoundland species, *nordenskioldii*, and the most modified in both is *nevadensis*, *lacustris* standing intermediate. The new form, again, is intermediate between *lacustris* and *nevadensis* proper and may be roughly characterized as uniting the characters of the fifth legs of the male and female and the caudal setæ of *nevadensis* with those of the abdomen of the male of *lacustris*.

---

* Herrick's species hardly seems to belong to this genus. The abdomen I understand was described as projecting from the left side of the abdomen, and consequently cannot be compared directly with that of *E. lacustris*, all of which are developed from the right; and the fifth feet of the male are not readily capable of close comparison with the corresponding appendages of any closed type. The singular once reported in the position of the hinge in the antennæ of the male also points to a difference of fundamental distinction, not easy to trace, and it is at best difficult to connect *nevadensis*, and *lacustris* with respect to the inner ramus of the swimming legs, the fifth legs of the male, and the caudal setæ.

It differs from typical *neradensis* in the more complete segmentation of the cephalothorax in both male and female, four sutures extending distinctly across the back. It is also a little larger, adult females measuring 2.12 to 2.4 millimeters in length to the tips of the rami, with an ordinary width across the cephalothorax of 6.4 millimeters. The male is somewhat smaller, about 2.1 millimeters in length.

The antennæ of the female are long and slender, reaching about to the posterior end of the penultimate segment of the abdomen. The first segment of the female abdomen is as long as the two following together, and the furca is as long as the preceding segment. The female abdomen is not curved as in *lacustris*, and the spermatophore extends downward and backward instead of curving upward, as in that species. The three caudal setæ are all similar and of equal width, the base of each being a third the width of the end of the furca. There is a short, stout, conical spine at the outer distal angle of each ramus, and a small, soft seta at the inner angle.

The fifth legs of the female are broader in proportion to their length than in *lacustris*, but more slender than in the *neradensis* of Lilljeborg. The last segment of the leg is four times as long as broad and bears six teeth (occasionally seven), four of which are terminal. The inner of these four is commonly the largest, although the third from within may equal it. The inner lateral tooth is close to the inner terminal, and nearly or quite equals it in size, and the outer lateral is nearly opposite. Sometimes there are two teeth on the outer margin of this segment. The middle joint of this leg is less than half as wide as long, and the basal is longer than wide.

In the male abdomen there are five distinguishable segments, as in all the other species,[*] the second, third, and fifth bearing lateral processes extending to the right. The first three segments are subequal in length. The lateral process of the second has the form of a stout but thin wing or lamina projecting laterally a distance equal to the width of the segment. It springs from a broad and thick prominence of the segment itself; is acute at the apex, with the point a little recurved, convex and smooth in front, as seen from above, and nearly straight behind, except that this edge is irregularly and minutely serrate throughout and deeply emarginate where it joins the segment. As seen from the side this blade is strongly curved downward (ventrally), like the following. The third segment bears a broad, thin lamina which projects outward and a little backward from its posterior angle as a flat process, as wide as long, curved downward and broadly rounded at the end, quite simple, except that it is strengthened beneath by a ridge of chitin. These processes are in strong contrast to the corresponding ones of *neradensis* proper. From the fourth segment spring two processes, the ventral of which is very similar to that of *neradensis*, but broader, a triangle in form, with nearly equal sides, with the apex slightly truncate and bearing three serrations, and with the posterior side very minutely roughened. The dorsal process of this segment is a small irregular plate curving forward, inward, and downward.

Fifth pair of legs substantially as in *neradensis* Lillj.

Abundant in Swan and Flathead lakes, Montana.

---

[*] A recent study of the male abdomen in *lacustris* shows beyond a doubt that the fourth and fifth segments are flexibly articulated and that the fourth is without process, the fifth bearing two processes, as in all the other species.

ROTIFERA.

Monostyla ovata, n. sp.

Lorica broad ovate, truncate, antennæ nearly as broad before as behind. Dorsal outline regularly rounded, not recurved before. Lobe flat; ventral plate flat. Toe with very distinct shoulder, however viewed. Front margins of both dorsal and ventral plates entire. Dorsal plate strengthened by two diverging longitudinal ribs, about equidistant from each other and from the lateral angles of the plate, rendering the anterior margin slightly angulate where they join it, and vanishing behind at about the middle of the shell. Eye single, transverse, oval, red, situated just above and before the mastax, with two very minute red points behind it. Foot and toe about two-thirds as long as the lorica is wide. Ventral plate much shorter posteriorly than dorsal, its posterior margin slightly excavated before the foot.

Dimensions, 0.25 millimeter long by 0.18 millimeter wide.

From warm spring (107° F.), Yellowstone Lake, August 3, 1890.

Conochilus leptopus, n. sp.

Resembling *C. volvox*. Antennæ adnate to the tip, where the pair are rounded off as one; but very slightly bifid at the base of the two hairs. Stalk not swollen, slender, tapering backward regularly from the slightly dilated point of attachment to the body. Eyes black, about midway between cloacal opening and edge of disk. Cloaca about half of distance from the edge of the disk to the base of the expanded body. Trophi slightly tinted yellowish brown, not orange.

Entire length, when expanded, 0.32 millimeter; extended stem, 0.13 millimeter; breadth of body, 0.088 millimeter; expanded disk, 0.094 millimeter.

Exceedingly abundant in spherical colonies in Yellowstone Lake, July and August, 1890, and also in Lewis Lake, Yellowstone Park, July, 1890.

PROTOZOA.

Stentor igneus, var. fuliginosus.

Form trumpate, with slightly swollen sides, very slightly changeable, peristome with spiral lobe, greatest width slightly more than half greatest length. Color soot black, given by blackish granules; when decolorized showing green as if by chlorophyll.

Individuals when highly magnified gray by reflected or transmitted light, through mingling of green and black.

Form not symmetrical, the right side (when spiral lobe of peristome is uppermost) being swollen below. When the peristomal lobe is at the side it appears as a tubercle or projection. Form sometimes considerably shortened, so as not to be longer than broad. Peristome angularly produced, so as to form with tip of spiral lobe an equilateral triangle, giving the entire animal an angular or trigonal shape. Tip flexible and contractile, transparent when extended.

Swarmed in Lake of the Woods August 20, 1890, forming soot-like collections as a film on the surface among pond lilies (*Nelumbium*) and a discontinuous coating on under sides of the same.

EXPLANATION OF PLATES

PLATE XXXVII

Fig. 1. *Daphnia pulex*, var. *pulicaria*. Male.
Fig. 2. *Daphnia dentifera.*
Figs. 3, 4. *Cypris barbatus.*

PLATE XXXVIII

Figs. 5-8. *Cypris barbatus.* (5) First antenna. (6) Second antenna. (7) Postabdomen. (8) Mandible.

PLATE XXXIX.

Figs. 9-12. *Cyclops thomasi.* (9) First leg. (10) Second leg. (11) Third leg. (12) Fourth leg.

PLATE XL.

Fig. 13. *Cyclops thomasi*, fifth leg.
Figs. 14-17. *Cyclops capilliferus.* (14) Adult female. (15) First leg. (16) Fourth leg. (17) Fifth leg.

PLATE XLI.

Fig. 18. *Cyclops capilliferus*, antenna of female.
Figs. 19-21. *Epischura nevadensis*, var. *columbiæ*. (19) Abdomen of male. (20) Fifth pair of legs of male. (21) Fifth pair of legs of female.
Fig. 22. *Diaptomus piscinæ*, fifth pair of legs of male.

PLATE XLII.

Figs. 23-25. *Diaptomus shoshone.* (23) Fifth pair of legs of male. (24) Fifth leg of female. (25) Right antenna of male.
Figs. 26-28. *Diaptomus lintoni.* (26) Fifth leg of female. (27) Fifth pair of legs of male. (28) Terminal segments of antenna.

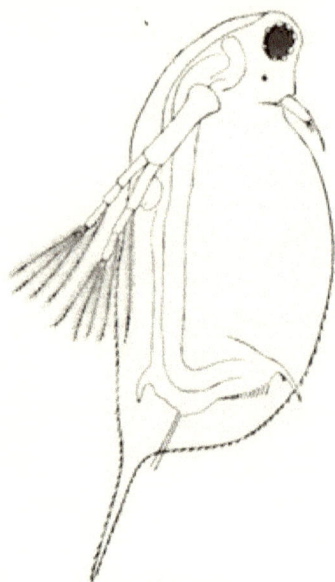

Fig. 1.

Fig. 2.

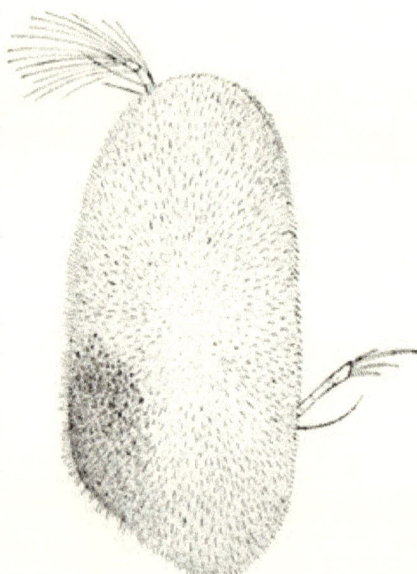

Fig. 3.

Fig. 4.

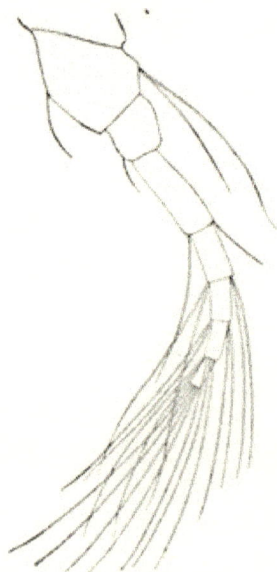

Fig. 5.

Fig. 6.

Fig. 7.

Fig. 8.

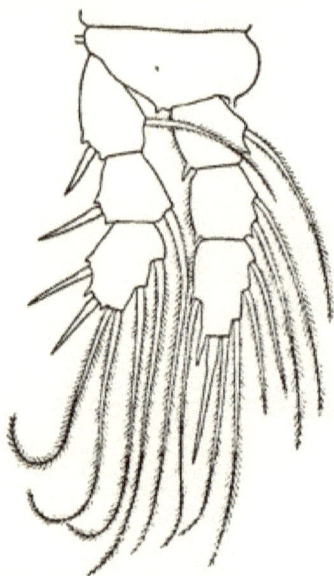

Fig. 9.

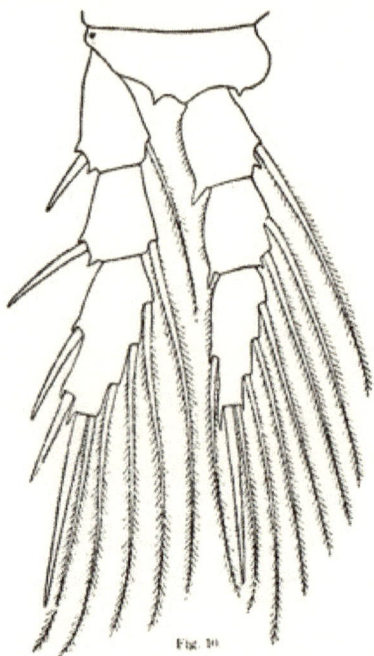

Fig. 10.

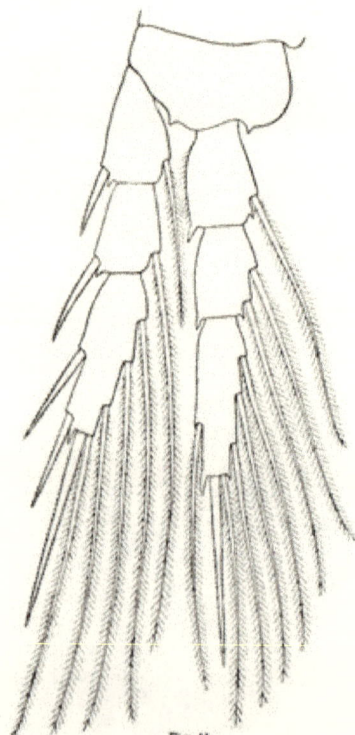

Fig. 11.

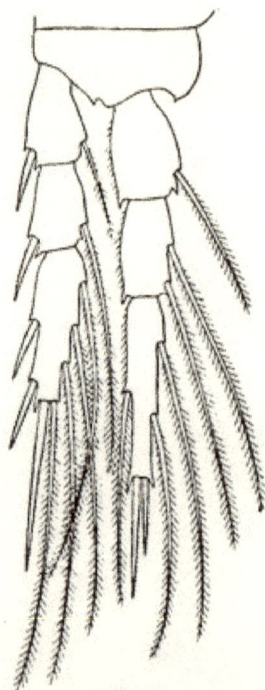

Fig. 12.

PLATE XL.

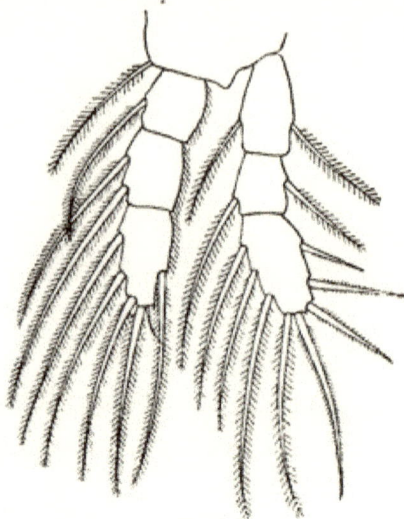

Fig. 15.

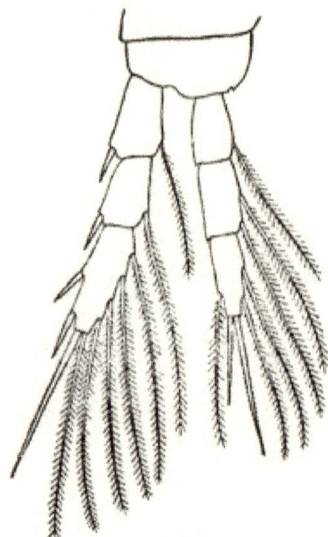

Fig. 16.

Fig. 13.

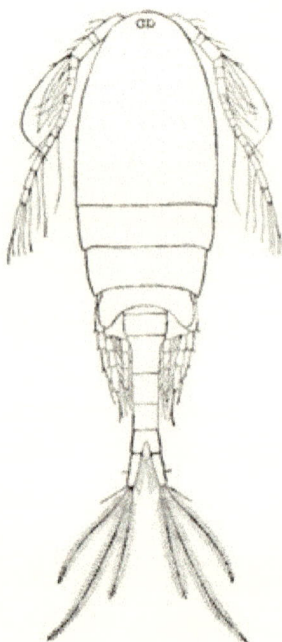

Fig. 14.

Fig. 17.

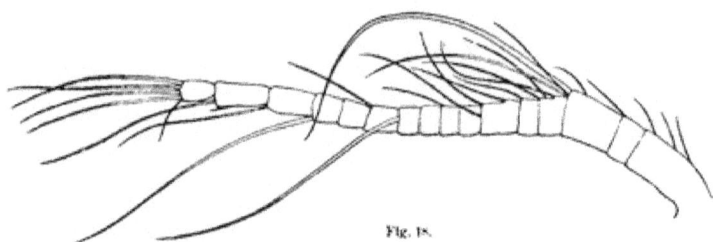

Fig. 18.

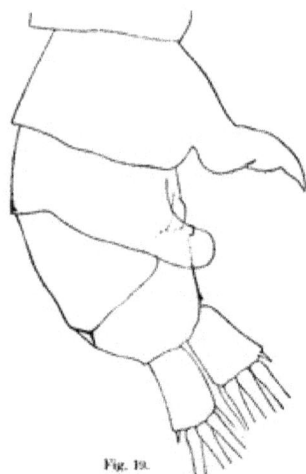

Fig. 19.

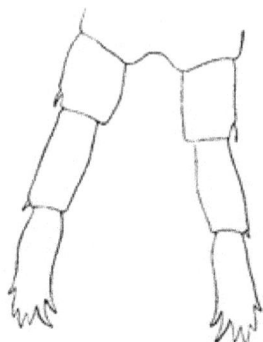

Fig. 21.

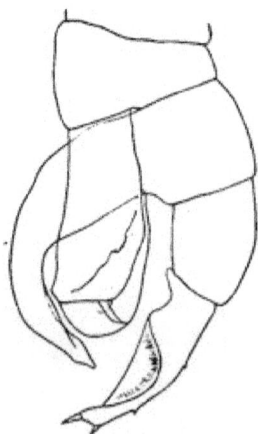

Fig. 20.

Fig. 22.

Bull. U. S. F. C. 1891. Aquatic Invertebrate Fauna of Wyoming and Montana. (To face page 258.)

PLATE XLII.

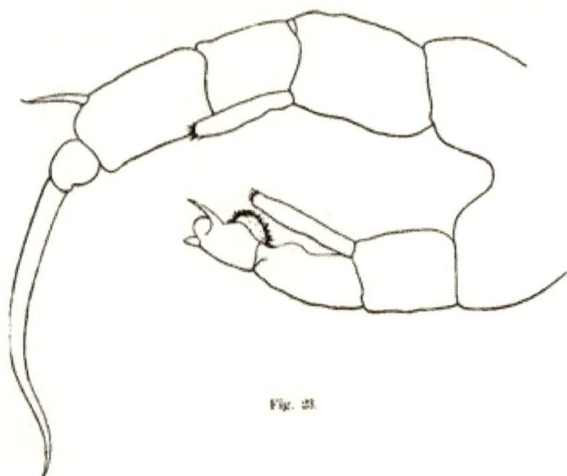

Fig. 23.

Fig. 24.

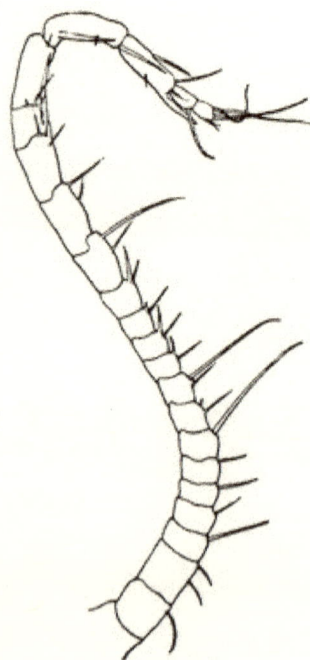

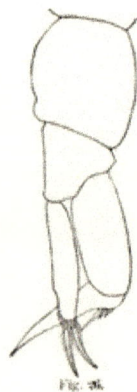

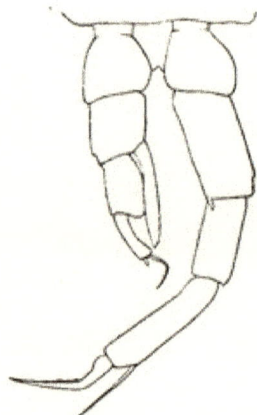

Fig. 26.

Fig. 27.

Fig. 28.

Fig. 25.

INDEX.

www.ingramcontent.com/pod-product-compliance
Lightning Source LLC
Chambersburg PA
CBHW020242090426
42735CB00010B/1804